The PIC Microcontroller:
Your Personal Introductory Course

Third edition

John Morton

ELSEVIER
BUTTERWORTH
HEINEMANN

AMSTERDAM • BOSTON • HEIDELBERG • LONDON • NEW YORK • OXFORD
PARIS • SAN DIEGO • SAN FRANCISCO • SINGAPORE • SYDNEY • TOKYO

Newnes is an imprint of Elsevier

Newnes

Newnes is an imprint of Elsevier
Linacre House, Jordan Hill, Oxford OX2 8DP, UK
30 Corporate Drive, Suite 400, Burlington, MA 01803, USA

First edition 1998
Second edition 2001
Third edition 2005
Reprinted 2006

British Library Cataloguing in Publication Data
A catalogue record for this book is available from the British Library

ISBN–13: 978-0-7506-6664-0
ISBN–10: 0-7506-6664-1

For information on all Newnes publications
visit our website at www.newnespress.com

Printed and bound in *Great Britain*

06 07 08 09 10   10 9 8 7 6 5 4 3 2

Contents

Acknowledgements

Max Horsey, Head of Electronics at Radley College in Abingdon and a great driving force for technological advancement, first introduced me to PIC micro-controllers in 1995. With the help of Philip Clayton I was shown a new concept in circuit design which opened up the possibility of new and more elaborate electronic devices.

I would like to take this opportunity to thank all those who have contributed, directly or indirectly, to make this book possible. First I must thank Richard Morgan, Warden of Radley College, for persuading me to try and get published, and my parents for their continual support with it. Chris, my brother, was an invaluable proof-reader and I must also thank Pear Vardhanabhuti who started out with no knowledge of programming, and bravely took on the task of learning all about PIC microcontrollers using just the book. He then went on to design and build the 'diamond brooch' project circuit board. Also helping to build projects were Ed Brocklebank, James Bentley and Matt Fearn, and Matt Harrison helped me with the artwork involved. My work was greatly facilitated by Philip Clayton, an immaculate technical proof-reader and advisor. Finally comes the most important thanks of all, to Max Horsey – a constant provider of assistance and advice, and fountain of new ideas; he has helped me immeasurably.

Preface to the third edition

When I was asked to write a new edition, I carefully read through the book trying to find how the current edition could possibly be improved. It was clearly a case of where to begin! With the help of several readers and their helpful emails, I have ironed out most of the, shall we say, elaborate spelling mistakes. My thanks therefore to Robert Czarnek, Lane Hinkle, Neil Callaghan, John Wrighte and Jimmy Gwinutt.

Since the first edition was published, I have received a great number of emails from readers asking for help with their various PIC projects. I am happy to help, and will try to answer any questions you may have. However, I have also been sent PIC programs without a single comment on them, and often without any indication of what task they are actually meant to perform, with a short message along the lines of: 'It doesn't work.' One of my favourite emails informed me that an error 'of type 0034q . 0089' kept occurring, and could I please fix it. These types of emails will seldom meet with a favourable response, simply because I haven't a clue what to do. So please *put comments everywhere* in your programs, and try to isolate exactly what is going wrong.

One of the major changes in this edition is the replacement of older one-time-programmable PIC microcontrollers with newer Flash versions. These are more suited to the kind of prototyping and testing that will take place as you go through the programs in this book, and develop on your own, as each PIC microcontroller can be programmed many times. These new PIC models can also be programmed *in-circuit*, so you don't even need to remove the PIC microcontroller from your board when updating the program. A short section introducing more advanced techniques, such as serial communication, has also been added to extend the scope of the book.

This book has been updated to conform to Microchip's trademark guidelines regarding the use of the word 'PIC'. PIC is a registered trademark of Microchip Technology Inc. in the US and other countries, and as such it should only be used as an adjective followed by an appropriate noun, such as 'PIC microcontroller'. If I have missed any instances of a lone 'PIC' without a suitable noun, please read it to yourself as 'PIC microcontroller'!

A final thanks must go to Max Horsey and the Electronics Department at Radley College who appear unaware that I have left the college, and continue to offer me use of their excellent facilities.

1
Introduction

It has now become possible to program microchips; gone are the days when circuits are built around chips, now we can build chips around circuits. This technology knows no bounds and complex circuits can be made many times smaller through the use of these *microcontrollers*, of which the PIC® is an excellent example. There is, however, little point in using a PIC microcontroller for a simple circuit that would, in fact, be cheaper and smaller *without* one. However, most complicated logic circuits could benefit immensely from the use of PIC microcontrollers. Furthermore, prototyping can be greatly enhanced as it's often much easier to make changes to a PIC program, than it is to start changing circuit designs and electronic components.

When you buy a PIC microcontroller, you get a useless lump of silicon with amazing potential. It will do nothing without – but almost anything with – the program that you write. Under your guidance, almost any number or combination of normal logic chips can be squeezed into one PIC program and thus in turn, into one PIC microcontroller. Figure 1.1 shows the steps in developing a PIC program.

PIC programming is all to do with numbers, whether binary, decimal or hexadecimal (base 16; this will be explained later). The trick to programming lies in making the chip perform the designated task by the simple movement and processing of numbers.

What's more, there is a specific set of tasks you can perform on the numbers – these are known as instructions. The program uses simple, general instructions, and also more complicated ones which do more specific jobs. The chip will step through these instructions one by one, performing millions every second (this depends on the frequency of the oscillator it is connected to) and in this way perform its job. The numbers in the PIC microcontroller can be:

1. **Received** from inputs (using an input 'port')
2. **Stored** in special compartments inside the chip (these are called 'file registers')
3. **Processed** (e.g. added, subtracted, ANDed, etc.)
4. **Sent out** through outputs (using an output 'port')

That is essentially all there is to PIC programming ('great' you may be thinking) but fortunately there are certain other useful functions that the PIC microcontroller provides us with such as an on-board timer (e.g. TMR0) or certain flags which indicate whether or not something particular has happened, which make life a lot easier.

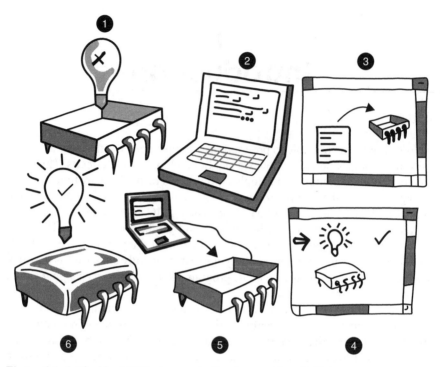

Figure 1.1 *1. The blank PIC microcontroller does nothing; 2. Write a program on a computer; 3. Pretend to program the PIC microcontroller on a computer; 4. Test the program on a computer; 5. Program a real PIC microcontroller; 6. Test the PIC microcontroller in a real circuit.*

The first chapter of this book will teach you how to use the PIC16F54 and 57. These are two fairly simple devices and knowledge of how to use them will serve as a solid foundation to move on from, as there are many other diverse and exciting PIC microcontrollers around, and indeed new ones coming out all the time. Subsequent chapters will introduce more advanced techniques, using the small 8-pin PIC12F508 and the versatile PIC12F675.

Some tips before starting

For those not familiar with programming at all, there may be some ideas which are quite new, and indeed some aspects of the PIC microcontroller may seem strange. Some of the fundamental points are now explained.

Binary, decimal and hexadecimal

First there is the business of different numbering systems: binary, decimal and hexadecimal. A binary number is a *base 2* number (i.e. there are only two types of digit (0 and 1)) as opposed to decimal – *base 10* – with 10 different digits

Table 1.1

Binary (8 digit)	Decimal (3 digit)	Hexadecimal (2 digit)
00000000	000	00
00000001	001	01
00000010	002	02
00000011	003	03
00000100	004	04
00000101	005	05
00000110	006	06
00000111	007	07
00001000	008	08
00001001	009	09
00001010	010	0A
00001011	011	0B
00001100	012	0C
00001101	013	0D
00001110	014	0E
00001111	015	0F
00010000	016	10
00010001	017	11
etc.		

(0 to 9). Likewise hexadecimal represents *base 16* so it has 16 different digits (0, 1, 2, 3, 4, 5, 6, 7, 8, 9, A, B, C, D, E and F). Table 1.1 shows how to count using the different systems.

The binary digit (or *bit*) furthest to the right is known as the least significant bit or *lsb* and also as *bit 0* (the reason the numbering starts from 0 and not from 1 will soon become clear). Bit 0 shows the number of 1s in the number. One equals 2^0. The bit to its left (*bit 1*) represents the number of 2s, the next one (*bit 2*) shows the number of 4s and so on. Notice how $2 = 2^1$ and $4 = 2^2$, so the bit number corresponds to the power of two which that bit represents, but note that the numbering goes from right to left (this is very often forgotten!). A sequence of 8 bits is known as a byte. The highest number bit in a binary word (e.g. bit 7 in the case of a byte) is known as the most significant bit (*msb*).

So to work out a decimal number in binary you could look for the largest power of two that is smaller than that number (e.g. 32 which equals 2^5 or $128 = 2^7$), and work your way down.

Example 1.1 Work out the binary equivalent of the decimal number 75.

Largest power of two less than $75 = 64 = 2^6$. Bit 6 = **1**
This leaves $75 - 64 = 11$ 32 is greater than 11 so bit 5 = **0**
16 is greater than 11 so bit 4 = **0**
8 is less than 11 so bit 3 = **1**

This leaves $11 - 8 = 3$ 4 is greater than 3 so bit 2 = **0**

2 is less than 3 so bit 1 = **1**

This leaves $3 - 2 = 1$ 1 equals 1 so bit 0 = **1**

So **1001011** is the binary equivalent.

There is however an alternative (and more subtle) method which you may find easier. Take the decimal number you want to convert and divide it by two. If there is a remainder of one (i.e. it was an odd number), write down a one. Then divide the result and do the same writing the remainder to the *left* of the previous value, until you end up dividing one by two, leaving a one.

Example 1.2 Work out the binary equivalent of the decimal number 75.

Divide 75 by two.	Leaves 37, remainder **1**
Divide 37 by two.	Leaves 18, remainder **1**
Divide 18 by two.	Leaves 9, remainder **0**
Divide 9 by two.	Leaves 4, remainder **1**
Divide 4 by two.	Leaves 2, remainder **0**
Divide 2 by two.	Leaves 1, remainder **0**
Divide 1 by two.	Leaves 0, remainder **1**

So **1001011** is the binary equivalent.

Exercise 1.1 Find the binary equivalent of the decimal number 234.

Exercise 1.2 Find the binary equivalent of the decimal number 157.

Likewise, bit 0 of a hexadecimal is the number of ones ($16^0 = 1$) and bit 1 is the number of 16s ($16^1 = 16$), etc. To convert decimal to hexadecimal (it is often abbreviated to just 'hex') look at how many 16s there are in the number, and how many ones.

Example 1.3 Convert the decimal number 59 into hexadecimal. There are three 16s in 59, leaving $59 - 48 = 11$. So bit 1 is 3. 11 is B in hexadecimal, so bit 0 is B. The number is therefore **3B**.

Exercise 1.3 Find the hexadecimal equivalent of 234.

Exercise 1.4 Find the hexadecimal equivalent of 157.

One of the useful things about hexadecimal is that it translates easily with binary. If you break up a binary number into four-bit groups (called *nibbles*, i.e.

small bytes), these little groups can individually be translated into one 'hex' digit.

Example 1.4 Convert 01101001 into hex. Divide the number into nibbles: 0110 and 1001. It is easy to see 0110 translates as 4 + 2 = 6 and 1001 is 8 + 1 = 9. So the 8 bit number is **69** in hexadecimal. As you can see, this is much more straightforward than with decimal, which is why hexadecimal is more commonly used.

Exercise 1.5 Convert 11101010 into a hexadecimal number.

An 8-bit system

The PIC microcontroller is an 8-bit system, so it deals with numbers 8 bits long. The binary number 11111111 is the largest 8-bit number and equals 255 in decimal and FF in hex (work it out!). With PIC programming, different notations are used to specify different numbering systems (the decimal number 11111111 is very different from the binary number 11111111)! A binary number is shown like this: **b'00101000'**, a decimal number like this: **d'72'**, or like this: .72 (it looks like 72 hundredths but it can be a lot quicker to write, if you use decimal numbers a lot). The hexadecimal numbering system is default, but for clarity write a small h after the number (the computer will still understand it and it reminds *you* that the number is in hex), e.g. 28**h**. Alternatively, you can write 0x at the start of the number (e.g. **0x3D**).

When dealing with the inputs and outputs of a PIC microcontroller, binary is always used, with each input or output pin corresponding to a particular bit. A **1** corresponds to what is known as *logic 1*, meaning the pin of the PIC microcontroller is at the supply voltage (e.g. +5 V). A 0 shows that pin is at *logic 0*, or 0 V. When used as inputs, the boundary between reading a logic 0 and a logic 1 is half of the supply voltage (e.g. +2.5 V).

Finally, if at any stage you wish to look up what a particular instruction means, refer to Appendix C which lists all of them with their functions.

Initial steps

The basic process in developing a PIC program consists of five steps:

1. **Select** a PIC model, and construct a program **flowchart**.
2. **Write** program (using Notepad provided with Microsoft Windows, or some other suitable development software).
3. **Assemble** program (changes what you've written into something a PIC microcontroller will understand).
4. **Simulate** or **emulate** the program to see whether or not it works.

5. **'Blow'** or **'fuse'** the PIC microcontroller. This feeds the program you've written into the actual PIC microcontroller.

Let's look at some of these in more detail.

Choosing your PIC microcontroller

Before beginning to write the program, it is a very good idea to perform some preliminary tasks. First you need some sort of project brief – what are you going to make and what exactly must it do. The next step is to draw a circuit diagram, looking in particular at the PIC microcontroller's inputs and outputs. Each PIC model has a specific number of inputs and outputs, you should use this as one of the deciding factors on which device to use and thus you should make a list of all the inputs and outputs required. In this book, we will abbreviate the full names PIC16F54 and PIC16F57 to 'PIC54' and 'PIC57', for the sake of brevity. The PIC54 has up to 12 input/output pins (i.e. it has 12 pins which can be used as inputs *or* outputs), and the PIC57 has up to 20.

Example 1.5 The brief is 'design a device to count the number of times a push button is pressed and display the value on a single seven-segment display. When the value reaches nine it resets.'

1. The seven-segment display requires **seven** outputs.
2. The push button requires **one** input, creating a total of 8 input/output pins. In this case a PIC54 would therefore be used (see Figure 1.2).

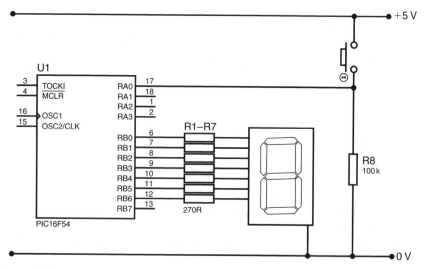

Figure 1.2

Make sure you employ **strobing** where possible. This is particularly useful when using more than one seven-segment display, or when having to test many buttons. Example 1.6 demonstrates it best:

Example 1.6 The brief is 'to design a system to test 16 push buttons and display the number of the button pressed (e.g. button number 11) on two seven-segment displays'.

It would first appear that quite a few inputs and outputs are necessary:

1. The two seven-segment displays require seven outputs each, thus a total of **14**.
2. The push buttons require one input each. Creating a total of **16**.

The overall total is therefore 30 input/output pins, which exceeds the maximum for PIC57. There are bigger PIC microcontrollers, with more than 30 pins, however it would be unnecessary to use them as this value can be cut significantly.

By strobing the buttons, they can all be read using only 8 pins, and the two seven-segment displays controlled by only 9. This creates a total of 17 input/output (or I/O) pins, which is under 20. Figure 1.3 shows how it is done.

By making the pin labelled RC0 logic 1 (+5 V) and RC1 to RC3 logic 0 (0 V), switches 13 to 16 are enabled. They can then be tested individually by examining pins RC4 to RC7. Thus by making RC0 to RC3 logic 1 one by one, all the buttons can be examined individually.

Strobing seven-segment displays basically involves displaying a number on one display for a short while, and then turning that display off while you display another number on another display. RB0 to RB6 contain the seven-segment code for both displays, and by making RA0 or RA1 logic 1, you can turn the individual displays on. So the displays are in fact flashing on and off at high speed, giving the impression that they are constantly on. The programming requirements of such a setup will be examined at a later stage.

Exercise 1.6 Work out which PIC model (PIC54 or PIC57) you would use for a device which would count the number of times a push button has been pressed and display the value on four seven-segment displays (i.e. will count up to 9999).

After you have selected a particular PIC model, the next step is to create a program flowchart (Example 1.7). This forms the backbone of a program, and it is much easier to write a program from a flowchart than from scratch.

A flowchart should show the fundamental steps that the PIC microcontroller must perform, showing a clear program structure. A program can have *jumps*, whereby as the PIC microcontroller is stepping through the program line by line, rather than executing the next instruction, it jumps to another part of the program. All programs require some sort of jump, as all programs must loop – they cannot just end.

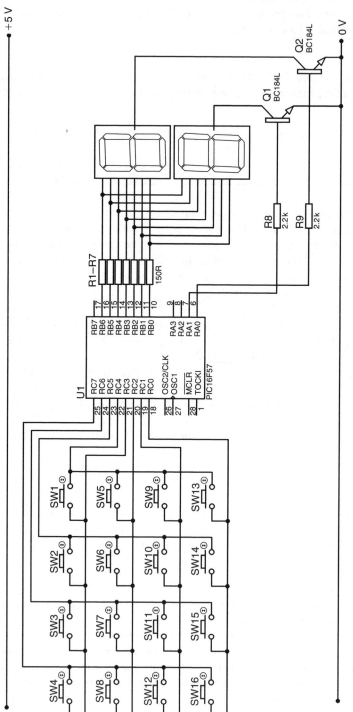

Figure 1.3

Example 1.7 The flowchart for a program to simply keep an LED turned on.

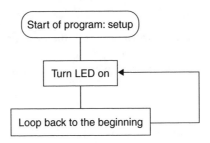

Figure 1.4

The setup box represents some steps which must be taken as part of the start of every program, in order to set up various functions – this will be examined later. Rectangles with rounded corners should be used for start and finish boxes.

Conditional jumps (in diamond shaped boxes) can also be used: *if* something happens, *then* jump somewhere.

Example 1.8 The flowchart for a program to turn an LED on *when* a button is being pressed.

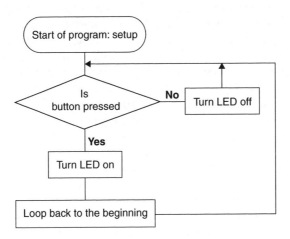

Figure 1.5

Sometimes a flowchart box may represent only one instruction, but sometimes it may represent a great deal, and such a diagram allows you to *visualise* the structure of your program without getting bogged down with all the nitty gritty instructions. Writing a program from a flowchart merely involves writing the

instructions to perform the tasks dictated by each box, and in this way a poten-
tially large program is broken down into bite-sized chunks.

Exercise 1.7 Draw the flowchart to represent the program required to make an
LED flash on and off every second (i.e. on for a second, then off for a second),
and a buzzer to sound for one second every five seconds.

Writing

Once the flowchart is complete, you should load up a PIC program template on
your computer (soon you will be shown how to create a sample template) and
write your program on it. All this can be done with a simple text program such
as Notepad, which comes with Microsoft Windows (or another suitable development package such as PIC PRESS – see Chapter 6).

Assembling

When you have finished writing your program, it is ready to be *assembled*. This
converts what you've written (consisting mostly of words) into a series of numbers
which the computer understands and will be able to use to finally 'blow' the PIC
microcontroller. This new program consisting solely of numbers is called the *hex
code* or *hex file* – a hex file will have **.hex** after its name. Basically, the 'complicated' PIC language that you will soon learn is simply there to make program writing easier; all a *raw* program consists of is numbers (some people actually write
programs using just numbers but this is definitely *not* advisable as it is a nightmare
to fix should problems arise). So the *assembler*, a piece of software which comes
with the PICSTART or MPLab package – called MPASM (DOS version) or
WinASM (Windows version) – translates your words into numbers. If, however, it
fails to recognise one of your 'words' then it will register an **error** – things which
are *definitely* wrong. It may register a **warning** which is something which is *probably* wrong (i.e. definitely unusual but not necessarily incorrect). The only other
thing it may give you is a **message** – something which *isn't* wrong, but shows it
has had to 'think' a little bit more than usual when 'translating' that particular line.
Don't worry if you are still a little confused by assembling, as all this will be
revised as you go through the process of actually assembling your program.

This assembled program will get fused into the *program memory*, when you
'blow' the PIC microcontroller. The PIC microcontrollers used in this book have
a Flash program memory, which can be re-written over and over again. Other
models may be OTP (one-time programmable), or UV-erasable.

You should now be ready to begin writing your first program …

The file registers

The key to the PIC microcontroller are its file registers. If you understand these
you're half way there. Imagine the PIC microcontroller as a filing cabinet, with

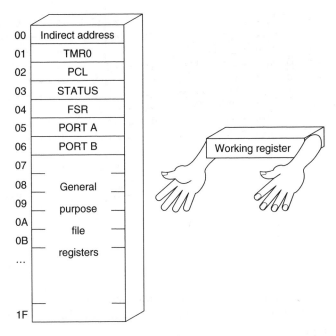

Figure 1.6 *Map of file registers for PIC16F54.*

many drawers, each containing an 8 bit number (a byte). These drawers are the file registers. As well as these file registers there is the *working register*. This register is different because it is not part of the filing cabinet. It is needed because only one drawer (i.e. file register) may be open at one time. So imagine transferring a number from one drawer to another. First, you open the first drawer, take the number out then close it, now … where is the number? The answer is that it is in the working register, a sort of bridge between the two file registers (think of it as the poor chap who has to stand in front of the filing cabinet). The number is temporarily held there until the second drawer is opened, upon which it is put away.

As you can see from Figure 1.6, each file register is assigned a particular number. You should call the file registers by their actual name when writing your program (as it is *much* easier to follow), and then the assembler will translate your names back to numbers when creating the hex file.

Do not worry about the names or functions of these file registers, they will be discussed later on. However, to summarise, registers 00 to 06 have specific functions, and registers 07 to 1F are *general purpose* file registers, which you have complete control over. You can use general purpose file registers to store numbers and can give them whatever name you want. Naturally you will need to tell the assembler how to translate your own particular names into numbers. For example, if you were to use file register 0C to store the number of hours that have passed, you would probably want to call it something like **Hours**. However, as the assembler is running through your program, it will not

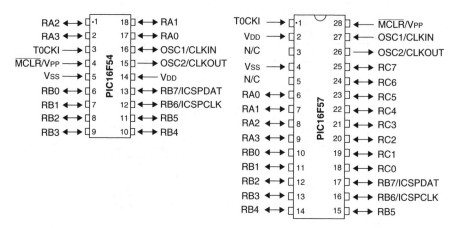

Figure 1.7

understand what you meant by 'Hours' unless you first *declare* it. You will be shown how and where to declare your file registers shortly, when we look at a program template.

Before this, a brief introduction to registers 05 and 06 is required ...

The *ports* are the connections between the PIC microcontroller and the outside world, its inputs and its outputs. The first port, Port A, has only 4 bits, i.e. it holds a nibble rather than a full byte and is the only register that does so. Each bit corresponds to a particular I/O (input/output) pin, so bit 0 of Port A corresponds to the pins labelled RA0 (pin 17 on PIC54 and 6 on PIC5 (Figure 1.7)). So when you write an 8-bit number into Port A, the four *most significant* bits are ignored, and likewise when you read an 8 bit number from Port A, the four most significant bits are read as 0.

For example, let us say that RA0, RA1, RA2 and RA3 are acting as inputs and there is a push button between each input and $+5$ V. If these push buttons are all pressed, the decimal number 15 (binary number 1111) would be in Port A. Conversely, if they are acting as outputs and are all connected to LEDs which were tied down to 0 V (as shown in Figure 1.9), moving the number 15 into Port A would turn all four LEDs on.

Exercise 1.8 Considering the arrangement just mentioned, in order to create a chase of the four LEDs (see Figure 1.8), a series of numbers will have to be moved into Port A one after another. What will these numbers be (answers in binary, decimal or hexadecimal)?

Port B (and Port C on PIC57) is simply another input/output port, just like Port A in all respects except that they have 8 bits (i.e. hold a byte). Port C on PIC57 is register 07, so note that the general purpose registers on this device start from 08 onwards.

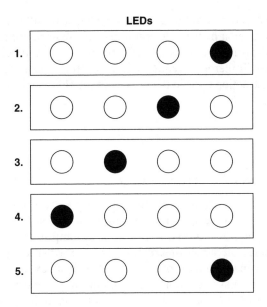

Figure 1.8

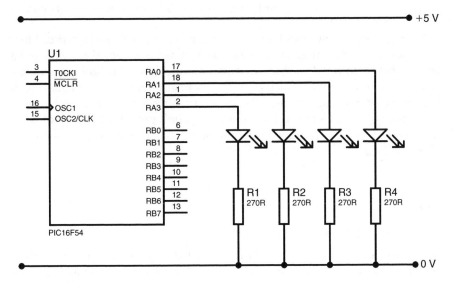

Figure 1.9

A program template

In this and subsequent sections you will begin to look at instructions. You may well find them unfamiliar, but fortunately there are a few general rules you can use to decipher an unknown instruction. First, wherever you come across the letter **f** in an instruction, it refers to a file register. A **w** will nearly always mean working register, and a **b** stands for **b**it in the vast majority of cases. Finally, an **l** will usually stand for **l**iteral, which effectively means number. An instruction containing an **l** will therefore require a number to be specified afterwards. For example, the instruction used in the next example (**bsf**) sets a bit in a file register (makes it 1).

Example 1.9

(Label) bsf   porta, 0   ; turns on LED

There are a few fundamental elements to writing a PIC program, one of these is line structure. Example 1.9 shows a sample line of programming. Optional first is a label which is required if you want to jump to this place in the program. Then comes the actual instruction: **bsf**, i.e. what are you *doing*. Third comes what are you doing it *to* (**porta, 0**), and lastly an explanation in your own words of what you have just done. It is important to note that you can write whatever you want in a PIC program as long as it is *after* a semicolon. Otherwise the assembler will try and translate what you've written (e.g. 'turns on LED') and will naturally fail and give you an **ERROR**. As the assembler scans through line by line, it will jump down to the next line once it comes to a semicolon.

I cannot stress how important it is to *explain* every line you write. First, what you've written may make sense as you write it, but there is a good chance that when you come back to it after a while, it will be difficult to understand. Secondly, it allows another person to read through your program with reasonable ease. It can sometimes be quite difficult to write a *good* explanation, as it should be very clear yet not too long. Don't get into the habit of basically copying out an instruction definition as your explanation, as shown in Example 1.10.

Example 1.10

bsf   porta, 0   ; sets bit 0 of Port A

The above comment means very little at all (it is easy to see that bit 0 is being set). It is far better to say *why* you have written what you have, and what its implications are (as shown in Example 1.9).

Now let's look at a program template, bear in mind this is simply an example and you may want to add or remove headings for your own personal template. In general, with your whole program, it is a good idea to space things out, and divide relevant sections up with lines. I suggest creating these with equal signs ($=$), of course you need a semicolon at the start of such a line.

Program template

```
;***********************************
; written by:                    *
; date:                          *
; version:                       *
; file saved as:                 *
; for PIC...                     *
; clock frequency:               *
;***********************************

; PROGRAM FUNCTION: ———————————————————————
;————————————————————————————————————————————


            list        P = 16F5x
            include     "c:\pic\p16f5x.inc"


;============
; Declarations:
porta    equ       05
portb    equ       06
(portc   equ       07)

            org         1FFh
            goto        Start
            org         0
;============
; Subroutines:
Init     clrf      porta          ; resets input/output ports
            clrf      portb
            (clrf     portc)
            movlw     b'xxxx'        ; sets up which pins are inputs and which
            tris      porta          ;   are outputs
            movlw     b'xxxxxxxx'
            tris      portb
            (movlw    b'xxxxxxxx'
            tris      portc)
            retlw     0
;============
; Program Start:
Start
            call    Init
Main
    (Write your program here)
            END
```

In the little box made up out of asterisks (purely there to make it look nice), there are a couple of headings which allow another reader to quickly get an idea of your program. Where it has: **for PIC...**, insert a model number such as 16F54 or 16F57, depending on which PIC you are using.

The clock frequency shows the frequency of the oscillator (*resistor/capacitor* or *crystal*) that you have connected. The PIC microcontroller needs a steady signal to tell it when to move on to the next instruction (in fact it performs an instruction every *four* clock cycles), so if, for example, you have connected a 4 MHz oscillator – i.e. four million signals per second – the PIC microcontroller will execute one million instructions per second. The clock frequency would in this case be 4 MHz.

Much more important than these headings are the actual preliminary *actions* that must be performed. The line: **list P = 16F5x** is incomplete. Replace the **5x** with the number PIC microcontroller you are using (e.g. 54), so a sample line would be: **list P = 16F54**. This tells the assembler which PIC microcontroller you are using.

The line: **include "c:\pic\p16f5x.inc"** enables the assembler to load what is known as a *look-up* file. This is like a translator dictionary for the assembler. The assembler will understand most of the terms you write, but it may need to *look up* the translations of others. All the file registers with specific functions (00 to 07) are declared in the look-up file. When you install PIC software it will automatically create these look-up files and put them in a directory (e.g. "C:/Program Files/Microchip/MPASM Suite/"). I have suggested you copy relevant look-up files (**.inc**) into a folder called "pic" in your C: drive so that it is easier to remember the correct path, but this is up to you. Regardless, you must write a valid path to the look-up file.

Next comes the space for you to make your *declarations*. These are, in a sense, your additions to the translator dictionary. If you were to declare **Hours** as file register **0C**, you would write the following:

```
;=============
; Declarations:
        Hours    equ    0Ch
```

You may also want to *re-declare* certain file registers with specific functions. This is because the assembler may be sensitive to whether something is upper case or in lower case. For example, the look-up file declares file register 05 as **PORTA**. Personally, I prefer writing it as **porta**, because it is quicker (I understand you may be happy to leave it as PORTA, but this example demonstrates the principle), so I will re-declare 05 as **porta** along with my other declarations:

```
;=============
; Declarations:
        porta    equ    05h
        Hours    equ    0Ch
```

This means I can write **porta** *or* **PORTA** and the assembler will understand *both* as file register 05. I also suggest declaring in order of increasing file register number.

Below the declarations are three lines which ensure the chip runs the program starting from the section labelled **start**. To understand this principle you must understand that every *instruction* line (i.e. not just a space or a line with some comments) has a particular number (or *address*) assigned to it.

Example 1.11

start

0043	**bsf**	**porta, 0**	**; turns on LED**
			; (This is to prove comments aren't counted)
0044	**goto**	**start**	**; loops back to start**

Notice how only the lines with instructions have addresses (**start** is merely a label and not an instruction). Now, the allocation of addresses is systematic – counting up as you go down the program – *unless* you tell it otherwise. You can actually label the next line with a particular address, and then the ones which follow will continue counting up from there. This is done with the assembler command **org**, followed by the address number you wish to give the next line.

Example 1.12

start

0043	**bsf**	**porta, 0**	**; turns on LED**
	org	**3**	**; makes the address number of the next**
			; instruction 3
0003	**bsf**	**porta, 1**	**; turns on buzzer**
0004	**goto**	**start**	**; loops back to start**

Notice how the command **org** is *not* given an address. This is because it is not an instruction which the PIC microcontroller executes, rather it is a note for the assembler telling it to stick the following instruction at (e.g.) address 0003 in the PIC microcontroller's program memory. Example 1.12 however would never work, because after executing address 0043, the chip would attempt to execute address 0044, but regardless it demonstrates the principle of the **org** instruction.

The PIC54 has 512 addresses (200h in hexadecimal) in its program memory, in other words it can hold programs which are up to 512 instructions in length. The first instruction to be executed when the PIC microcontroller is switched on (or reset) is called the *reset vector*, and points to address 1FFh for the PIC54. We want the PIC microcontroller to begin at the place in the program which we have labelled start, so we make sure the instruction at 1FFh is **goto   start**. In the template, **org** is used to place instruction **goto   start** at 1FFh, making it the first to be executed. However, subsequent instructions must start counting from 0, so the following command is **org   0**. Writing the program memory address by the instructions shows how it works:

```
              org     1FF
01FF          goto    start
              org     0
;===========
; Subroutines:
0000   Init   clrf    porta   ;
0001          clrf    portb   ;
etc.
```

The first instruction to be executed (**goto   start**) makes the chip **goto** (*jump*) to the part of the program labelled **start**, and thus the PIC microcontroller will begin running the program from where you have written **start**. Different PIC models have different reset vectors (it's 7FFh for the P16F57), so the program template should be changed accordingly.

The next section of the template holds the *subroutines*. These are quite complicated and will be investigated at a later stage; all you need know at the moment is that the section labelled **Init** is a subroutine, and it is accessed using the **call** instruction. The subroutine **Init** should be used to set up all the particulars of the PIC microcontroller. With the PIC5x series of chips, this mainly involves selecting which pins of the PIC microcontroller are to act as inputs, and which as outputs. In other cases with more complex PIC models, more setting up will be required. Please note that this setting up is put in the **Init** subroutine only to get it out of the way of the main body of the program and thus make it neater and more reader friendly. First we use the instruction:

```
clrf      FileReg   ;
```

This **clear**s (makes zero) the number in a file register. We use it at the start of the setup subroutine to make sure the ports are reset at the start of the program. This is because after the PIC microcontroller is reset, the states of the outputs are the same as they were before the reset. However, in some cases where you want the states of the ports to be retained from before the reset, these clearing instructions may need to be removed. If the PIC model that you're using doesn't contain a Port C, do not bother clearing it.

The next instruction is:

```
movlw   number   ;
```

It **mov**es the **l**iteral (the **number** which follows the instruction – in the first case **b'xxxx'**) into the working register. Then the instruction **tris** takes the number in the working register and uses it to select which bits of the port are to act as inputs and which as outputs. A binary **1** will correspond to an input and a **0** corresponds to an output. Pins which you don't use are best made outputs.

Example 1.13 Using a PIC54, pins RA0, RA1 and RA3 are connected to push buttons. Pins RB0 to RB6 are connected to a seven-segment display, and pins RA2 and RB7 are connected to buzzers. What should you write to correctly specify the I/O pins?

```
movlw   b'1011'
tris    porta
movlw   b'00000000'
tris    portb
retlw   0
```

There are two things to notice: first, there is no specification of Port C (naturally as the PIC54 doesn't have one), and secondly, a reminder that bit numbering goes from right to left (it is easy to forget!).

Exercise 1.9 Using a PIC57, pins RA1 and RA2 drive LEDs, pins RA0 and RA3 are connected to temperature sensors, RB0 to RB6 control a separate chip, and RB7 is connected to a push button. RC1 to RC5 carry signals to the PIC microcontroller from a computer, and all other pins are not connected. What should you write in the **Init** section of the program?

The instruction **retlw** is placed at the end of a subroutine, normally with a **0** after it.

Finally the last part of the template holds **Start**, where the program begins. Notice that the first thing that is done is setting up the ports' inputs and outputs. After the line **call Init**, there is the heading **Main** after which you write your program. At the end of your program, you must write **END**.

2
Exploring the PIC5x series

Your first program

For this chapter (and subsequent ones) it is assumed you are sitting in front of a computer which has the application Notepad or PIC PRESS (see Chapter 6). Do not worry if you don't have any actual PIC software at the moment, as the programs you write now can be assembled later, when you do actually get some software.

If using Notepad, you should start by copying out a program template; save the file as **template.asm** and make sure you select **any file** as the file type. The **.asm** shows that the file is an *assembly source*, i.e. it is something to be assembled, which makes it recognisable to the assembler. To begin with we'll be using the PIC54, so make the necessary alterations on the template (from now on do *not* simply **Save**, but instead **Save As**, so the file **template.asm** remains unchanged). Call this new file **ledon.asm**.

The first program you will write will be very simple. It simply turns on an LED (and keeps it on indefinitely). This will simply use two instructions: **bsf** and **goto**.

The instruction **bsf** sets (i.e. makes 1), a particular **b**it in a **f**ile register. You therefore need to specify the file register and the bit after the instruction (what you are doing it *to*).

Example 2.1 **bsf   portb, 5   ; turns on buzzer**

portb is the file register, and **5** is the number of the bit being set. There is a comma between the file register and the bit.

You should already be familiar with the instruction **goto label** (remember **goto start** from the template?). It makes the PIC microcontroller jump to the section of the program you have labelled **label**. Naturally you can name the place to which you want it to jump anything you want, but it is a good idea to make it relevant to what is going on in the program in that particular section. Be careful, however, not to give sections the same name as you give to general purpose file registers, otherwise the assembler will get confused.

The first step of writing a program is assigning inputs and outputs. For this device we simply need one output for the LED. This will be connected to RA0 (pin 17) of the PIC microcontroller. The second step is the program flowchart shown in Figure 2.1.

We can now write the program. You should be able to set up the inputs and outputs yourself (remember if a pin is not connected, make it an output). You can also have a go at writing the program yourself (it should consist of two lines).

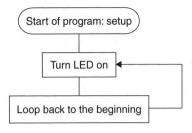

Figure 2.1

The first box (Set up) is performed in the Init subroutine. The second box involves turning on the LED. This involves making RA0 high (+5 V), and thus bit 0 of Port A should be 1 (i.e. *set*). To do this we use the instruction **bsf.** The line after . . .

Start   call   Init   ;

. . . should therefore be:

Main   bsf   porta, 0   ; turn on LED

Remember, a program cannot just end; it must keep looping, so the next box involves making the program jump back to the beginning. The next line should therefore be:

goto   Main   ; loops back to Main

Note that it should *not* go back to **Start**, as this will do the setting up all over again. Depending on how you wrote Port A in the program, you may need to redefine it in the declarations section. This would be necessary unless you wrote **PORTA** (i.e. in upper case).

The program is now ready to be assembled and you may want to check you have everything correct by looking at the program in its entirety. This (along with all the other example programs) is shown in the program section in Chapter 7. This program has been given the name *Program A*.

We now turn to assembling the program. You can download assemblers from a variety of sources or use the built-in assembler in PIC Press. I will discuss a popular development environment from Microchip (the makers of PIC microcontroller) called MPLab, which can be downloaded from www.microchip.com. The discussion refers to MPLab IDE v7.00, but the steps described are unlikely to change significantly for future versions.

Open MPLab IDE, select File → Open and find your assembly file (e.g. ledon.asm). This should create a window containing your assembly file, with basic colour coding. Assembler commands (such as **org** and **equ**) appear in blue plaintext, while PIC instructions (such as **clrf** and **goto**) appear in blue bold.

Before assembling your code, you should select the PIC model you're using in Configure → Select Device. To assemble your source file, go to Project → Quickbuild *filename*.asm (where *filename* should be the name of your source file). An Output window will appear summarising any Errors, Warnings or Messages. If there are any errors (or warnings you wish to change), note the line number on which they occur. To find the relevant line in the source file use CTRL + G to jump to a line number (also, the line number of the cursor is shown at the bottom of the screen). After you have assembled the file with no errors, a **.hex** file is loaded into the memory. You can use this file to simulate the program, and to blow the PIC microcontroller. To save this file, select File → Export . . . , click OK, and then type the name of your file. You should use the same name as your source file (e.g. ledon.hex).

It is worth noting that MPLab also comes with the standalone assembler, MPASMWIN, which you can use to assemble source files without loading MPLab. If you open this assembler, a window will appear with several parameters that need to be set. Click Browse . . . to select the file which you wish to assemble (the *Source file*). Leave all parameters at *Default*, and I would recommend selecting only the 'List File' under *Generated Files*. This list file is useful when it comes to tracking down the errors that you made in the source file (if any!). It lists the errors within your program, next to where they occur. You can open this file, and search for instances of the word 'error' to track down your errors. Alternatively, a really quick way to assemble is to drag the .asm file over MPASMWIN – this should start the assembly process.

Configuration bits

There are a handful of settings which are hard-wired into the PIC microcontroller when it is programmed, called 'configuration bits'. The number and type of these bits vary for different models, but for the PIC54 we have the following:

Code Protect:	On *or* Off
Watchdog Timer:	On *or* Off
Oscillator Selection:	LP *or* XT *or* HS *or* RC

'Code protect' is a feature which prohibits the reading of a program from the PIC microcontroller. For testing purposes, it is best to turn this feature *off*. The watchdog timer is discussed on page 69, but until then we should turn it *off*. Finally, the oscillator selection tells the PIC microcontroller what kind of oscillator you plan to connect (these are described in the next section). These features can be selected using tick boxes at the programming stage, but they can also be specified in the program using the __**config** command (note this has two underscore characters at the start). For example, to disable code protect and the watchdog timer, and to select the crystal oscillator, we would write:

__config   _CP_OFF & _WDT_OFF & _XT_OSC

The exact words for each feature (e.g. **_WDT_OFF**) can be found in the include file for the relevant PIC model. Separate each feature with an ampersand (&).

Testing the program

In general, there are three steps to testing a program:

1. **Simulating**
2. **Emulating**
3. **Blowing** a PIC microcontroller and putting it in a circuit

Simulating

The first of these, simulating, is entirely software based. You simply see numbers changing on the computer screen and need to interpret this as whether or not the program is working. Select Debugger → Select Tool → MPLab SIM to activate the simulator. Assuming you have loaded a source file in MPLab and assembled it, your program should be loaded into the memory ready for use by the simulator. Press F6 (or Debugger → Reset → Processor Reset) to reset the program. A green arrow should appear at the line **goto   Start** indicating that this is the next instruction to be executed. Press F7 (Debugger → Step Into) to execute an instruction one step at a time. The first time you press F7 the green arrow should jump to **call   Init**. Continue *stepping* through your program and you will see the flow of the program, eventually ending up in the final loop.

In order to faithfully simulate the behaviour of the final PIC microcontroller, the simulator requires that the configuration bits are correctly defined. This is done through Configure → Configuration Bits . . . and ticking the appropriate boxes.

We now wish to see how the registers of the PIC microcontroller (and in particular its outputs) are changing throughout the program. Go to View → Special Function Registers to load a window showing the states of PIC registers (presented as binary, decimal and hexadecimal). Reset the program back to **goto   Start**, but this time look at the special function registers (in particular PORTA and PORTB) as you step through the program. After passing through the **Init** subroutine, PORTA should be set to 0. Then you can see the line starting **bsf ...** turn on bit 0 of PORTA (in other words, making pin RA0 high). We will return to the simulator later to see how to set the states of inputs, for programs that respond to external stimuli.

Emulating

A more visual (but much more expensive) step in testing employs an emulator (such as *PICMASTER* from Microchip and *ICEPIC* from RF Solutions). These use a probe in the shape of a PIC microcontroller which comes from your PC and plugs into a circuit board. You can then load and run your program, much like simulating, with the great advantage that the program responds to the states of the inputs of the probe, and the pins of the probe change according to the program flow. This not only presents a more visual demonstration of the program, but allows you to test both the program *and* its implementation in a real circuit.

Blowing the PIC microcontroller

The final step involves actually putting the program into the PIC microcontroller. You should only do this once you have tested the program, either through simulation or in-circuit emulation. In order to do this, you need a PIC programmer, and circuit board in which to place your chip after it's programmed. There are a great many programmers available, though ones which are compatible with MPLab allow for a seamless transition from the steps described above, to the final programming step. Such programmers include *PICStart Plus* (from Microchip) and *PIC MCP* (from Olimex).[1] Note that although third-party alternatives may appear more inexpensive, the documentation can sometimes leave a little to be desired, so they may not be appropriate for the true novice.

In-circuit serial programming (ICSP) allows the transfer of a program to a PIC microcontroller, while it remains in its own circuit board. The *Baseline Flash Microcontroller Programmer* (BFMP) is a very handy ICSP tool for the PIC16F54, PIC12F508 and PIC12F675 which are used in the example projects of this book, as well as a number of other PIC models. It is a compact module with a USB interface to your PC, which can plug into your custom circuit-board to program the PIC microcontroller and provide power.

The *PICKit™ 1 Flash Start Kit* is a development board which supports similar devices. It is also a USB device, and interfaces either with MPLab or with a piece of standalone programming software (called *PICkit™ 1*). It even comes with a PIC12F675 ready for you try out the projects in Chapter 4. The board comes with 8 LEDs, a button, and a variable resistor connected. You can either use these components as they are on the board (shown in Appendix H), or use a jumper cable to connect to your own board. You can keep the PIC microcontroller on the development board, and use the jumper leads to connect to your own external components. However, it is also possible to use the jumper leads to go to a complete external board including the PIC microcontroller and use the board as an in-circuit serial programmer (but if you do this, make sure you keep the leads short). In either case, you need to take into account the components already attached to the pins on the PICKit board, as these may disrupt the intended behaviour.

Hardware

Figure 1.7 shows the pin arrangements for the PIC54 and PIC57. The pins labelled RAx, RBx, and RCx are I/O pins. V_{DD} and V_{SS} are the positive and 0 V supply pins respectively. The positive supply should be between 2.0 and 5.5 V, but note that the maximum operating frequency depends on the supply voltage. For

---

[1]See Appendix G: Contact Information and References for more information.

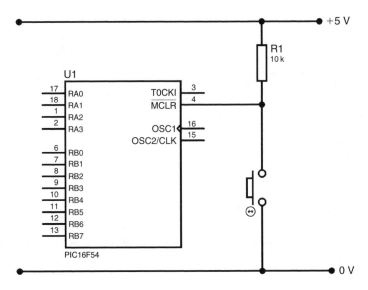

Figure 2.2

example, for a 2 V supply, the maximum operating frequency is 4 MHz (equivalent to 1 million instructions per second). Above 4.5 V, the maximum operating frequency is as high as 20 MHz (5 million instructions per second). The pin labelled T0CKI is the **Timer Zero Clock Input** – the PIC microcontroller can be set to automatically count signals on this pin. On older PIC models, this pin might be labelled RTCC (**Real Time Clock Counter**). $\overline{\text{MCLR}}$ is the **Master Clear** pin (a reset pin). The bar over the top means it is *active low*, in other words when you make this pin low (0 V), the PIC microcontroller drops what it's doing and returns to **goto   start** (or wherever the reset vector is pointing to). Figure 2.2 shows how to trigger the $\overline{\text{MCLR}}$ by means of a push button reset. The resistor is there to tie the $\overline{\text{MCLR}}$ high when the button is not being pressed.

In a real circuit, we require a short delay between the circuit first being powered up, and the program commencing. This is necessary since many power supplies take a short time to stabilise, and crystal oscillators also need a 'warm-up'. Many PIC microcontrollers (including the PIC54) therefore come with a *Device Reset Timer* (DRT), which provides a delay of approximately 18 ms by keeping the PIC microcontroller in a Reset condition for a short time after power is supplied. If the supply or oscillator is particularly unstable (requiring a longer delay), or the PIC model you are using does not have a DRT, you will need to attach a small circuit to the $\overline{\text{MCLR}}$, as shown in Figure 2.3. The value of C1 can be increased to lengthen the power-up delay.

The chip also needs a steady pulse to keep it going (an oscillator). This can be created using a crystal, or resistor/capacitor arrangement. The most accurate and reliable is likely to be a crystal oscillator, as it is less affected by external

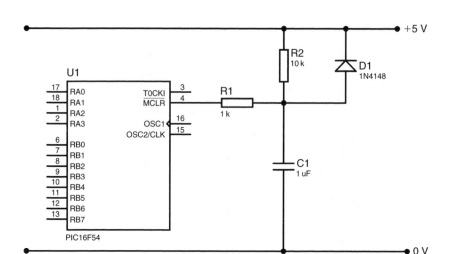

Figure 2.3

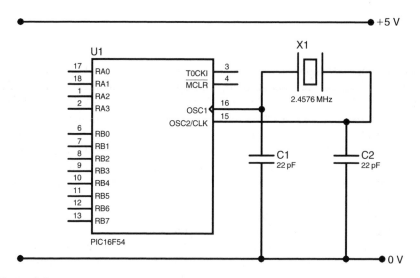

Figure 2.4

variables such as temperature. If you use a crystal, and desire high-speed operation, I recommend a 16 MHz crystal oscillator. For lower-speed operation,
2.4576 MHz is a convenient frequency. Also note that ceramic oscillators provide a smaller, lower-cost alternative to quartz crystals. Crystal oscillators
should be connected as shown in Figure 2.4 (though 10 pF capacitors should be
used for higher frequencies such as 16 MHz). Alternatively, you may want to
drive the PIC microcontroller from an external clock source, especially if you
want to synchronise two devices. To do this, simply connect the clock source to
the OSC1 pin (CLKIN). The oscillator frequency divided by 4 is available as a

Table 2.1

Cext	Rext	Average Fosc @ 5 V, 25°C	
20 pF	3.3 k	4.973 MHz	±27%
	5 k	3.82 MHz	±21%
	10 k	2.22 MHz	±21%
	100 k	262.15 kHz	±31%
100 pF	3.3 k	1.63 MHz	±13%
	5 k	1.19 MHz	±13%
	10 k	684.64 kHz	±18%
	100 k	71.56 kHz	±25%
300 pF	3.3 k	660 kHz	±10%
	5.0 k	484.1 kHz	±14%
	10 k	267.63 kHz	±15%
	160 k	29.44 kHz	±19%

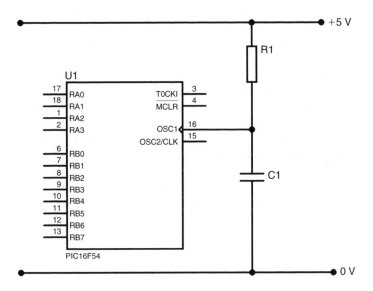

Figure 2.5

clock source for other devices from the OSC2/CLKOUT pin. Finally, while prospect of running at PIC microcontroller at high speed may appear attractive, remember that this consumes more power, and so should be avoided where unnecessary.

Resistor/capacitor oscillators are a good choice when accuracy and stability are not important. Useful values are shown in Table 2.1, while the appropriate arrangement is shown in Figure 2.5.

A circuit diagram for the LED ON project is shown in Figure 2.6 (I have chosen a resistor/capacitor oscillator as accurate timing in not required). The connections for an ICSP (in-circuit serial programmer) are also shown – these can be ignored if a separate programmer (such as *PICStart Plus*) is used.

If you aren't using ICSP, but are using a standalone programmer like the PICStart Plus, you can load the .asm file in MPLab, assemble it and make sure the configuration bits are correctly set. Then select the programmer you're using (Programmer → Select Programmer), enable it (Programmer → Enable Programmer), and then write the program to the PIC microcontroller (Programmer → Program). You can use the other options in this menu to erase chips (assuming they are electrically erasable), and read the program off the chip.

If you are using ICSP, the *Baseline Flash Microcontroller Programmer* (BFMP) is an ideal choice, and I would recommend including a socket which interfaces with this programmer (with the arrangement shown in Figure 2.6), on each of your development boards. Note the three connections to the PIC microcontroller: the VPP connection goes directly to the $\overline{\text{MCLR}}$ pin, which has a 10 k tie-up resistor to V_{DD}, and the ICSPDAT/CLK lines go directly to RB7/6. In this way, you can reprogram your PIC microcontroller without taking it out of the board. The BFMP uses programming software called PICkit™ 1 which is much more basic than MPLab – it just takes an assembled .hex file and writes it to the PIC microcontroller. You can't set the configuration bits in the software using tick boxes, but instead have to use the **_config** command (discussed previously) in your assembly code so that the configuration bits are part of the .hex file.

Connect the BFMP to your PC via a USB cable (note that when ordering a BFMP it is likely *not* to come with the USB cable), and connect it to your board through the 5 connector pins. The PICkit™ 1 programming software has two varieties: 'Classic' and 'Baseline Flash'. For programming the PIC16F54 and PIC12F508 models, use the 'Baseline Flash' software and select the correct PIC

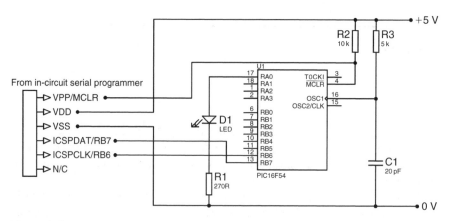

Figure 2.6

type in the drop-down box. In the programmer window, load the program (File →
Import Hex), and then press the **Write Device** button. A nice feature is that if
the .hex file changes (you make some changes to the program, and re-assemble),
it is automatically reloaded before programming the PIC microcontroller. If you
have any problems writing, try erasing the PIC microcontroller first by pressing
Erase. You can also use this software to read the program off a chip. After programming the PIC microcontroller, the LED connected to the RA0 pin should
now be on. All this just to see an LED turn on may seem like a bit of an anticli-
max, but there are greater things to come!

Using the testing instructions

A far more useful program would turn on an LED *if* a push button is pressed,
and then turn it off when it is released. This will involve *testing* the state of the
input pin connected to the push button. There are two basic methods of testing
inputs:

1. Testing a particular bit in the port, using the **btfss** or **btfsc** instructions.
2. Using the entire **number** held in the port's file register to look at all the
   inputs as a whole.

In most cases you tend to test particular bits, and as there is only one push button, only one bit will need to be tested. The push button will be connected to pin
RB0, and again the PIC54 will be used. Two I/O pins will be needed in this new
device, and the flowchart is shown in Figure 2.7. The circuit diagram is shown
in Figure 2.8. However, if you are using ICSP, add a 10 k resistor between the
MCLR pin and V$_{DD}$, as before.

Again, you should be familiar with the *set up* part, and be able to write it
yourself. The next box requires the use of the new instruction **btfss**. This
instruction tests a bit of a file register and will skip the next instruction if the bit

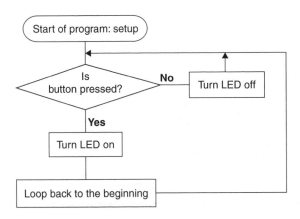

Figure 2.7

is set (i.e. if it is high or logic 1). Its 'sister' instruction is **btfsc** which again tests a bit of a file, but this time skips the next instruction if the bit is clear (i.e. if it is low or logic 0). So to test the push button, the instruction line is:

btfss   portb, 0   ; tests the push button

If the button pulls the input pin *high* when it is pressed, the program will execute the next instruction if the button is *not* pressed. In such a case the LED should be turned off and then the program should loop back to **Main**. The way to do this is to make the program *jump* to a section labelled something like **LEDoff**. This requires the instruction:

goto   LEDoff   ; jumps to the section labelled LEDoff

After this line is the instruction that will be executed if and only if the push button is pressed. This should therefore make the LED turn on. You should already know how to do this, as well as the instruction that follows it which makes the program loop back to **Main**. This leaves us with the section labelled LEDoff. In this section the LED should be turned off, and then the program should loop back to **Main**. To turn a bit off use the instruction **bcf**. This clears a bit of a file register and works just like **bsf**. The next line is:

LEDoff   bcf   porta, 0   ; turns off LED

We finally come to the last instruction which again should make the program loop back to **Main**. You should be able to do this yourself. The program is now ready to be assembled, but again you may check that the program is correct by looking at the whole program (named *Program B*). Load this program into MPLab, and assemble it as before. We will now simulate this program, but in order to do this

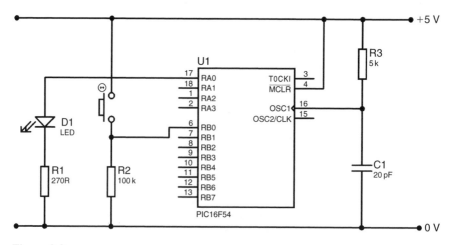

Figure 2.8

we need to simulate *inputs*. Activate the simulator, turn off the WDT, and open the window for the Special Function Registers. As you step through the program, you will see its behaviour for the case where the push button is not pressed. To tell the simulator that the button is pressed, open Debugger → Stimulus Controller → New Scenario. This lists a number of inputs that you can control manually. Click in one of the boxes marked 'Pin', and select RB0 (as the button is attached to RB0). For 'Action' select *Toggle*. Under 'Comments' you can type something like 'Push button' to remind you what this input represents. A little arrow should appear at the beginning of the line, under 'Fire'. Whenever you click the arrow, the state of RB0 will toggle. You won't see the effect immediately in the Special Function Registers window – you need to step through one instruction in order for the change to register. Use this to set RB0 high, and go through the program to check it works when the button is pressed (you should see PortA bit 0 go from 0 to 1).

As well as stepping through a program line by line, the simulator also allows us to run through the program at high speed. In order to tell it when to stop running, we need to set a *break point*. When the simulator reaches this point, it stops running and you can continue stepping through slowly. Set the push button to the off state (make sure RB0 is clear), and put your cursor on the line that turns the LED on. Right-click and select "Set Breakpoint". If you now tell the simulator to Run (F9, or Debugger → Run), it should never encounter your breakpoint, and will continue indefinitely. While it's running, click on the push button. The simulator hits your breakpoint and stops. Breakpoints are particularly useful when you wish to quickly go through a part of the program that you are not interested in (e.g. you already know it works) and go through a later section more slowly.

It turns out that the seven line program we wrote above to make a push button turn on an LED, is in fact very inefficient (the same task can be accomplished with only three lines)! You may be wondering how this can be, as we went through all the development steps and constructed a logical flow chart, but somehow there is a much better way.

Sometimes it helps to step back from the problem and look at it in a different light. Instead of looking at the button and LED as separate bits in the two ports, let's look at them with respect to how they affect the entire number in the ports. When the push button is pressed the number in Port B is **b'00000001'**, and in this case we want the LED to turn on (i.e. make the number in Port A **b'00000001'**). When the push button isn't pressed, Port B is **b'00000000'** and thus we want Port A to be **b'00000000'**. So instead of testing using the individual bits we are going to use the entire number held in the file register (think back to the two different testing methods introduced at the start of this section). The entire program merely involves moving the number that is in Port B into Port A. As you know this cannot be done directly and involves the moving of the number in Port B to the working register, and then moving the number from the working register into Port A. To move (in fact *copy*) the number from Port B into the working register we need the following instruction:

```
movf    FileReg, w      ;
```

This **mov**es the number from a file register into the **w**orking register. This instruction is very often abbreviated to:

> **movfw   FileReg   ;**

This instruction will do exactly the same thing, and is translated to the same number by the assembler. So the instruction to move the number from Port B into the working register is:

> **movfw   portb   ; moves the number in Port B to the**
> **   ; working reg.**

Then to move the number into Port A, we need the instruction:

> **movwf   FileReg   ;**

This **mov**es the number from the **w**orking register into a file register. To move the number from the working register into Port A we would write:

> **movwf   porta   ; moves the number from the w. reg. into**
> **   ; Port A**

After these two lines we need only loop back to **Main** so it cycles through these two lines constantly. Please note this shorter technique can only be used because the push button and LED are connected to the particular pins described in this example. Unless you specifically connect them up so that the technique works, it is unlikely to do so. This shorter program is shown as Program C in Chapter 7.

The circuit diagram for this project is the same as with the previous version, which is shown in Figure 2.8. The next section will introduce timing which is where the PIC microcontroller will really begin to get useful.

Timing

The PIC microcontroller comes with an on board timer called **TMR0** (in more advanced chips there is more than one timer, e.g. TMR1, TMR2, etc.). As you may remember, TMR0 (said timer zero) is file register number 01. It has two basic modes: counting an *internal* or *external* signal. When on the internal counting mode, the number it holds counts up at a constant rate (depending on the oscillator which you've attached). When counting external signals, it counts the number of signals received by the timer zero clock input (pin 3 on the PIC54 and 1 on the PIC57). When the number passes 255, it resets and continues from 0 again, as with any file register (this is called *rolling over*). As you can already see, there are various settings for the TMR0 and these can be controlled by the bits in the **OPTION** register. This register will *not* be familiar as it wasn't on the diagram showing the file registers (see Figure 1.6). This is because it isn't a file register that you can directly access (at least on the PIC5x series). In order to put a number into it, you first load the number into the working register, and then write the instruction: **option**. This automatically takes the number from the working register and moves it into the OPTION file register. The bits in the OPTION register are allocated as shown below.

This may be hard to follow, but this is basically how all file registers are explained in the PIC databook, so it is important to be familiar with the format. In the OPTION register each bit controls a particular setting, except bits 6 and 7. As you can see they have no purpose and are read as 0. Bit 5 (T0CS) is the *TMR0 clock source*, and defines whether TMR0 is counting internally (using the oscillator) or externally (counting signals on the T0CKI pin). Bit 4 (T0SE) selects the *TMR0 source edge*, and is fairly irrelevant if counting internally, but can be important if counting external signals. It selects whether TMR0 counts up every time a signal drops from logic 1 to logic 0 (i.e. *falling edge triggered*), or when the signal rises from logic 0 to logic 1 (i.e. *rising edge triggered*).

Bit no. 7	6	5	4	3	2	1	0	|TMR0	WDT
–	–	**T0CS**	**T0SE**	**PSA**	**PS2**	**PS1**	**PS0**	|**Rate**	**Rate**
0	**0**				0	0	0	|**1:2**	**1:1**
					0	0	1	|**1:4**	**1:2**
					0	1	0	|**1:8**	**1:4**
					0	1	1	|**1:16**	**1:8**
					1	0	0	|**1:32**	**1:16**
					1	0	1	|**1:64**	**1:32**
					1	1	0	|**1:128**	**1:64**
					1	1	1	|**1:256**	**1:128**

Prescaler assignment
0 – if you want the prescaler to be used by the TMR0
1 – if you want the prescaler to be used by the WDT

TMR0 source edge
0 – if you want to count up when the signal rises
1 – if you want to count up when the signal drops

TMR0 clock source
0 – if you want to count an internal signal
1 – if you want to count an external signal (on the T0CKI pin)

Now we come to the *prescaler* bits (PS2, PS1 and PS0). As you already know, the PIC microcontroller divides the frequency of the oscillations it receives from its oscillator (crystal, R/C, etc.) by four, and uses this as its driving frequency. This same value is used by TMR0 when counting internally. Let's take a typical oscillator frequency of 2.4576 MHz. This is divided by four leaving 0.6144 MHz, in other words a signal which oscillates 614400 times a second. When trying to use TMR0 to count seconds, minutes and even days, it is clear that a file register which counts up so fast is of little use. TMR0 would have to count up to 614400 for one second to pass, but of course it resets at 255 and would never reach this number. TMR0 has to be therefore *prescaled*, i.e. its frequency needs to be reduced. By the use of bits 0 to 2 in the OPTION register, TMR0 can automatically be prescaled by up to 256 times. When using TMR0 to count seconds and minutes, etc. it would be necessary to prescale it by the maximum amount. Prescaling TMR0 by 256 divides the frequency of 614400 Hz by 256, to 2400 Hz (surprisingly the numbers work out nicely!). So even with maximum prescaling, TMR0 still counts up once every 1/2400th of a second. We need to slow it down further ourselves, and this will be explained shortly.

The only bit left unexplained is bit 3 (PSA), the *prescaler assignment* bit. This introduces the idea of a *WDT* or *watch dog timer*, which is explained in a later section. This bit selects whether it is the WDT that is being prescaled or the TMR0 – you can only prescale one of them. Which ever one *isn't* being prescaled can still run, but with no reduction of the timer's frequency (i.e. a prescaling factor of 1).

Example 2.2 What number should be moved into the OPTION register in order to be able to use the TMR0 efficiently to eventually count the number of seconds which have passed?

Bits 6 and 7 are always 0.
TMR0 is counting *internally*, so bit 5 (T0CS) is 0.
It's irrelevant whether TMR0 is *rising* or *falling edge triggered* so bit 4 (T0SE) is 0 or 1 (let's say 0).
Prescaling for TMR0 is required, so bit 3 (PSA) is 0.
Maximum prescaling of 256 is required, so bits 2 to 0 (PS2-0) are all 1.

Hence the number to be moved into the OPTION register is: **00000111**.

Exercise 2.1 What number should be moved into the OPTION register in order to be able to use the TMR0 to count the number of times a push button is pressed?

Exercise 2.2 **Challenge!** What number should be moved into the OPTION register so that TMR0 can keep track of the number of times a push button is pressed, such that it resets when the maximum of 1023 presses is reached?

Now that you know what number to move into the OPTION register, you need to know *how* to move it. This calls for a familiar instruction: **movlw**. As you may remember, this moves the number that follows it into the working register. Then the instruction **option** moves the number from the working register into the OPTION register.

Example 2.3 **movlw   b'00000111' ; sets up TMR0 to count**
**   option   ; internally, prescaled by 256**

Notice how the explanation describes the *two* lines – rather than doing each one in turn, it makes sense to look at the instruction *pair*. As you are unlikely to want to keep changing the TMR0 settings it is a good idea to place the above instruction pair in the **Init** subroutine, to keep it out of the way.

If you want to be timing seconds and minutes, you need to perform some frequency dividing yourself. This is basically the same as prescaling, but as it takes place after the prescaling of TMR0, we should call it *postscaling*. This requires quite a complex instruction group, but let's try to build it up step by step. First, the essence of postscaling is counting the number of times a rising file register (like the TMR0) reaches a certain value. For example, we need to wait until the TMR0 counts up to 2400 times, for one second to pass. This is the same as waiting until the TMR0 reaches 30, for a total of 80 times, because $30 \times 80 = 2400$ (think about it).

How do we know when TMR0 has reached 30? We subtract 30 from it, and see whether or not the result is zero. If TMR0 *is* 30, then when we subtract 30 from it, the result will be zero. However, by subtracting 30 from the TMR0 we are changing it quite drastically, so we use the command:

subwf   FileReg, w

This **sub**tracts the number in the **w**orking register from the number in a **f**ile register. The **,w** after the specified file register indicates that the result is to be placed back in the **w**orking register, thus leaving the original file register number *unchanged*. In this way we can subtract 30 from TMR0, without actually changing the number in TMR0, i.e. see what *would* happen to TMR0 if we were to subtract 30.

The next problem is finding out whether or not the result of the operation mentioned above is zero. This is done using one of the PIC *flags* mentioned in Chapter 1. The flag we use is the *zero flag*. A flag is merely one bit in a register (number 02), which is automatically set or cleared depending on certain conditions. The zero flag is set when the result of an operation is zero, and is cleared when the result isn't zero. You already know the instruction for testing a bit in a file register, in this case the instruction line would be:

btfss   STATUS, Z   ; tests the zero flag (skip if the result was 0)

Rather than specifying the bit number after the file register, as is normally the case (e.g. **porta, 0**) – which in this case would be 2 – it is advisable to write **Z**, because it is understood by the assembler (with the help of a lookup file) and it is easier for you to understand. There are only a few select cases where this kind of substitution may be used.

So far, we have managed to work out when the TMR0 reaches the number 30. We need this to happen 80 times for one second to pass; this is best done using the following instruction line:

decfsz   FileReg, f

This will **dec**rement (subtract one from) a **fi**le register, and **s**kip the next instruction if the result is **z**ero. This is in effect a shortcut, and the identical operation could be performed over numerous steps, including the testing of the zero flag. Thus if the number in the specified file register is initially 80, the program will pass this line 80 times until it skips. If the next instruction is a looping instruction (i.e. one which makes the program jump back to the beginning of this timing section, the program will keep looping until the number in the file register reaches 0 (i.e. it will loop 80 times), after which it will skip the looping instruction and proceed onto the next part of the program. For this whole timing concept to work, the program must only execute this **decfsz** instruction when the TMR0 has advanced by 30 (e.g. gone from 0 to 30 *or* from 30 to 60, etc.). If we are in a looping system, it is all very well to test for TMR0 to reach 30 the first time round, but it will take another 256 advances of TMR0 to reach 30 for a second time (the TMR0 will continue counting up past 30, reset at 255, and then continue from 0). We could therefore reset TMR0 every time it reaches 30, but other parts of the program may be using it and would be relying on it counting up steadily and continuously. A better solution is to change the number you are waiting for TMR0 to reach. The second time round the loop it would be necessary to test for TMR0 to reach 60 (i.e. 30 + 30), and then the next time 90 (60 + 30), etc. The number we are testing for should therefore be held in a file register (let's call it **Mark30**, because it **mark**s when TMR0 has advanced by **30**), and every time the TMR0 'catches up' with **Mark30**, 30 must be added to it. The instruction pair for this involves a new instruction:

      addwf   FileReg, f   ;

This **add**s the number in the **w**orking register to the number in a **f**ile register, and leaves the result in the **f**ile register. So we need to move the number we want to add to the file register into the working register first. The required instruction pair to add the decimal number 30 to a file register called **Mark30** would therefore be:

      movlw   d'30'   **; adds 30 to Mark30**
      addwf   Mark30, f ;

When we need to access this number, it will be necessary to move (in fact *copy*) the number from the file register to the working register. As you know this involves the instruction **movfw**.

The file register which we are decrementing (which holds the number 80 to start with) shall be called **Post80** (Timer **Post**scaler by a factor of **80**).

The program section which follows is the entire instruction set required to create a one second delay. The first four lines where numbers are being moved into Mark30 and Post80 may be placed in the **Init** subroutine. Read through the instruction set carefully, we will be using this technique in the next example program. Please note that GPF stands for general purpose file register.

```
        movlw   d'30'           ; moves the decimal number 30 into
        movwf   Mark30          ;  the GPF called Mark30, the marker

        movlw   d'80'           ; moves the decimal number 80 into
        movwf   Post80          ;  the GPF called Post80, the first
                                ;  postscaler
TimeLoop
        movfw   Mark30          ; takes the number out of Mark30
        subwf   TMR0, w         ; subtracts this number from the
                                ;  number in TMR0, leaving the result
                                ;  in the working register (and leaving
                                ;  TMR0 unchanged)
        btfss   STATUS, Z       ; tests the zero flag – skip if set, i.e. if
                                ;  the result is zero it will skip the next
                                ;  instruction
        goto    TimeLoop        ; if the result isn't zero, it loops back to
                                ;  'TimeLoop'

        movlw   d'30'           ; moves the decimal number 30 into
        addwf   Mark30, f       ;  the working register and then
                                ;  adds it to Mark30

        decfsz  Post80, f       ; decrements Post80, and skips the next
                                ;  instruction if the result is zero
        goto    TimeLoop        ; if the result isn't zero, it loops back to
                                ;  'TimeLoop'

; When it reaches this point, 1 second has passed

        movlw   d'80'           ; resets Post80, moving the number 80
        movwf   Post80          ;  back into it
```

The next example project will be an LED which turns on and off every second and a buzzer which sounds for one second every five seconds. This will involve two outputs, one for the LED and one for the buzzer. The LED will be connected to RA0, and the buzzer to RB0. The oscillator should be accurate so a crystal arrangement will be used, running at 2.4576 MHz. The program flowchart for this project is shown in Figure 2.9, and the circuit diagram in Figure 2.10.

The set up should present no problems, remember to define any general purpose file registers such as Mark30 and Post80 using the **equ** instruction. You can make them file register numbers 08 and 09 for example. In the **Init** subroutine you may want to specify the number that goes in the OPTION register.

The instruction set for the whole of the box 'Wait one second' is the program section mentioned previously which creates a 1 second time delay. At the end of the section (the line after the **movwf** instruction), the state of the LED must be changed (if it is on, turn it off and vice versa). There are two methods of achieving

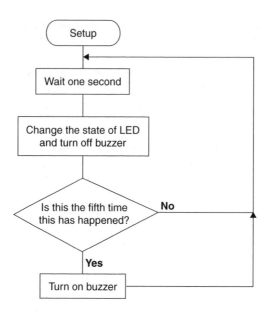

Figure 2.9

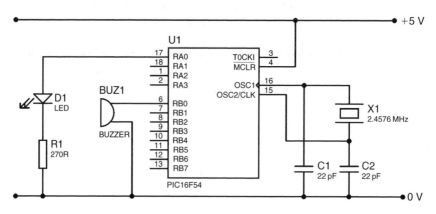

Figure 2.10

this. First, the current state of the LED can be tested (using the **btfss** or **btfsc** instructions), after which the program branches off to one of two sections depending on the LED's state, which will then either turn it on or off. Far easier when the rest of the I/O port is empty (there are no other connections to Port A apart from the LED), is to use the following instruction:

> **comf   FileReg, f   ;**

This instruction **com**plements (toggles the state of all the bits in) a file register, and leaves the result in the file register. We can use this because even though it will affect all the other bits in Port A (RA1, RA2 and RA3), this doesn't matter as they aren't connected to anything. To toggle the state of the bits in Port A the instruction would be:

comf   porta, f   ; toggles the state of the LED

However in most cases it won't be possible to simply toggle (change the state of) *all* the bits in a file register, so selective toggling must be carried out. This is done using the *exclusive OR* logic command. A logic command looks at one or more bits (as its inputs) and depending on their states produces an output bit (the result of the logic operation). The table showing the effect of the more common *inclusive OR* command on two bits (known as a *truth table*) is shown below.

Inputs		Result
0	0	0
0	1	1
1	0	1
1	1	1

The output bit (**result**) is high if either the first **or** the second input bit is high. The exclusive OR is different in that if *both* inputs are high, the output is low:

Inputs		Result
0	0	0
0	1	1
1	0	1
1	1	0

One of the useful effects is that if the second bit is 1, the first bit is toggled, and if the second bit is 0, the first bit isn't toggled (see for yourself in the table). In this way certain bits can selectively be toggled. If we just wanted to toggle bit 0 of a file register, we would exclusive OR the file register with the number **00000001**. This is done using one of the following instructions:

xorwf   FileReg, f   ;

This exclusive **OR**s the number in the working register with the number in a file register, and leaves the result in the file register. Each bit is exclusive ORed to each other according to bit number (bit 0 with bit 0, bit 1 to bit 1, etc.). Alternatively it may be more suitable to use:

xorlw   number   ;

This exclusive **OR**s the number in the working register with a literal (**number**).

Exercise 2.3 Two instructions are needed to toggle bits 3, 5, and 7 of Port B, what are these two lines?

The other task that must be completed is turning off the buzzer. Most of the time the buzzer won't have been on anyway, but for the one in five times that it *is* on, this will turn it off after one second has passed. This is done using the **bcf** instruction.

Finally we need to see if this is the fifth time one second has passed (i.e. have five seconds passed?). This is done, as before, using the **decfsz** instruction. Use another general purpose file register called **_5Second** (the underscore at the start of the name is there because a file register name cannot start with a number). The number 5 should be moved into it to begin with, and then after the **decfsz** instruction is reached five times, it will skip the next instruction, which should therefore be some sort of looping instruction. After the number reaches 0, and therefore five seconds have passed, the number 5 should be moved back into **_5Second**, because otherwise it will take another 256 seconds for the value to reach 0 again (as with the resetting of **Post80** in the previous example). When five seconds have passed, the buzzer should be turned on, and then program loops back to the beginning.

The whole program is shown in Program D; load it into MPLab and begin simulation. We would like to monitor the states of our general purpose registers (GPFs), namely **Mark30**, **Post80**, and **_5Second**. Click View → Watch, to open a window of *watch registers* – i.e. registers which you would like to monitor during simulation. Double click in a box under 'Symbol Name' and enter 'Mark30'. The simulator will recognise this as address 08, and also present its current value, in whatever format you request. To add new formats (e.g. binary, or decimal), right-click on one of the column titles, and select the desired format. Add the other two GPFs, load the Special Function Registers window, and then begin stepping through the program. During the **Init** subroutine you should see values entered into these registers. You then enter a loop where we wait for the TMR0 to count up. Set a *break point* immediately after the loop (tip: you can also do this by double-clicking a line), and press Run. When the simulator reaches the break, the TMR0 should have reached 30. 30 will then be added to Mark30, 1 will be subtracted from Post80, and the program will loop back – and you can watch all this happen in the relevant windows.

So far we have covered quite a few instructions and it is important to keep track of all of them, so you have them at your fingertips. Even if you can't remember the exact instruction name (you can look these up in Appendix C), you should be familiar with what instructions are available.

Exercise 2.4 What do the following do? **bsf**, **bcf**, **btfss**, **btfsc**, **movlw**, **movwf**, **movfw**, **decfsz**, **comf**, **subwf**, **addwf**, **equ**, **option**, **goto**, **tris**, **iorlw**, **iorwf**, **xorlw** and **xorwf**. (Answers in Appendix C.)

Explain also the significance of **,f** or **,w** after the specified file register, with certain instructions, such as **subwf**, **addwf**, **comf**, and **decfsz**, etc. (Answers in Appendix I.)

There will also be another example project using most of the ideas we have so far covered: a traffic lights system. There will be a set of traffic lights for

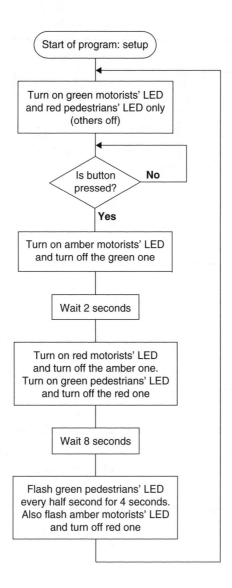

Figure 2.11

motorists (green, amber and red), and a set of lights for pedestrians (red and green), with a button for them to press when they want to cross. This makes a total of five outputs and one input, and thus the PIC54 will be used.

The red, amber and green motorists' lights (LEDs) will be connected to RB0, RB1 and RB2 respectively. The pedestrian push button shall go to RA0, with the red and green pedestrian lights to RB4 and RB5 respectively. The flowchart is shown in Figure 2.11, and the circuit diagram in Figure 2.12.

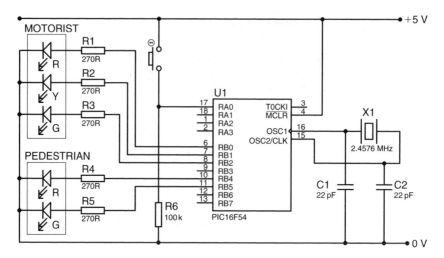

Figure 2.12

To start with, the motorists' light should be green, with all the others off, until the push button is pressed. The red pedestrian light should be on, and the green one off. All this should present no great problem, however rather than setting and clearing the individual bits, simply move the correct number into Port B.

Exercise 2.5 What *two* lines will be used to get the LEDs in the correct states?

There then needs to be a loop where the pedestrian's push button is tested continually, the program should only jump out of the loop when the button is pressed.

Exercise 2.6 What *two* lines will this loop consist of?

As soon as the button is pressed (i.e. after the loop is jumped out of) the amber motorists' light should be turned on, and the green one turned off. There should be no change to the pedestrians' lights.

Exercise 2.7 What *two* lines will accomplish these required output changes?

As the flowchart in Figure 2.11 shows, there are quite a few time delays required, and rather than copy the same thing over and over again for each time delay, it makes sense to use a time delay *subroutine*. Subroutines will be fully discussed in detail in the next section on seven-segment displays, however we will merely *use* one in this program as the general concept is simple. All we need know for the moment (and this should be familiar from studying the program template) is that when you access a subroutine, the program jumps to a certain place, runs through some instructions, and then returns to where it left of. To access a subroutine, the instruction is **call**, and to return to the line after the call instruction, you need to write **retlw**. This instruction must always be followed by a number, but in cases where this number is not important you can simply write **0** (as you may remember from the **Init** subroutine).

In this program, we will create a subroutine which creates a short delay. Whenever we want a delay to occur we can simply *call* the subroutine, and then know that after the required time has passed the program will return to where it left off. To be able to use the delay subroutine for all delays, the delay will have to be programmable from *outside* the subroutine. This delay subroutine will be just like the one-second time delay used previously, with the exception that we wish it to work for delays of 0.5, 2 or 8 seconds. We can therefore use a fixed *marker* of 240, and a variable *postscaler* of 5, 20, or 80 depending on what time we require. We can use the working register to carry the message to the subroutine, by moving the required postscaler value into the working register before the **call** command, and then moving the contents of the working register into a postscaler register *in the first line of the subroutine*. For a delay of 2 seconds, all we need to write in the body of the program is:

```
        movlw   d'20'      ; sends message of 2 seconds to sub
        call    TimeDelay  ; creates delay of required time
```

As long as the subroutine **TimeDelay** began as follows:

```
TimeDelay   movwf   PostX     ; sets up variable postscaler
            movlw   d'240'    ; sets up fixed marker
            movwf   Mark240   ;

TimeLoop    etc. (as previous time delays)
```

Exercise 2.8 Write the full **TimeDelay** subroutine. Don't forget to add the line **retlw   0** at the end of the subroutine.

After the two-second delay, the red motorists' light must be turned on, and the amber one off. The red pedestrian light must be turned off, and the green one turned on.

Exercise 2.9 What *two* lines will make the required output changes?

Exercise 2.10 Now an eight-second delay is required. What *two* lines will create the required delay?

Exercise 2.11 After the eight-second delay the red motorists' light should be replaced for the amber one. What *two* lines accomplish this?

Now both the green pedestrians' light and the amber motorists' light must flash on and off every half second for four seconds (the output should toggle every half second, eight times).

Exercise 2.12 **Challenge!** What *eight* lines will flash the lights as described? HINT: Think of a compact way to run a flashing loop eight times – you will need to use a general purpose file register.

The traffic lights now return to their original states, and the program can loop back to **Main**. You have basically written this whole program yourself; to check the entire program, look at Program E.

Seven-segment displays

Using a PIC microcontroller to control seven-segment displays allows you to display whatever you want on them. Obviously all the numbers can be displayed, but also most letters: A, b, c, C, d, E, F, h, H, i, I, J, l, L, n, o, O, P, r, S, t, u, U, and y.

The pins of the seven-segment display should all be connected to the same I/O port on the PIC microcontroller, in any order (this may make PCB design easier). The spare bit may be used for the dot on the display. Make a note of which segments (a, b, c, etc.) are connected to which bits.

Example 2.4 Port B Bit 7 = d, Bit 6 = a, Bit 5 = c, Bit 4 = g, Bit 3 = b, Bit 2 = f, and Bit 1 = e.

The number to be moved into Port B when something is to be displayed should be in the format **dacgbfe-** (it doesn't matter what bit 0 is as it isn't connected to the display), where each letter corresponds to the required state of the pin going to that particular segment.

The segments on a seven-segment display are labelled as shown in Figure 2.13.

So if you are using a common cathode display (i.e. make the segments high for them to turn on – see Figure 2.14), and you want to display (for example) the letter **P**, you would turn on segments a, b, e, f, and g.

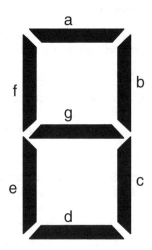

Figure 2.13

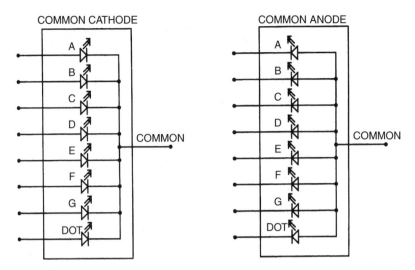

COMMON CATHODE   COMMON ANODE

Figure 2.14

Given the situation in Example 2.4, where the segments are arranged **dacgbfe-** along Port B, the number to be moved into Port B, to display a **P** would be **01011110**. Bit 0 has been made 0, as it is not connected to the display.

Example 2.5 If the segments of a common cathode display are arranged **dacgbfe-** along Port B, what number should be moved into Port B, to display the letter **I**, and the letter **C**?

The letter **I** requires only segments b and c (or e and f) so the number to be moved into Port B would be **00101000** or **00000110**.

The letter **C** requires segments a, d, e, and f, so the number to be moved into Port B would be **11000110**.

Exercise 2.13 If the segments are arranged **dacgbfe-** along Port B, what number should be moved into Port B to display the numbers 0, 1, 2, 3, 4, 5, 6, 7, 8, 9, A, b, c, d, E, and F?

This conversion process of a number into a seven-segment code can be carried out in various ways, but by far the simplest involves using a *subroutine*. To convert the number you want displayed into an actual display code, a decoding subroutine should be used. The general idea is that you first load the number to be displayed into the working register, then *call* the subroutine, which will then return to the program with the appropriate code in the working register.

Let's call the subroutine **_7SegDisp**, and store the number we want displayed in a file register called **Display**. The seven-segment display will be connected to

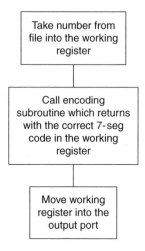

Figure 2.15

Port B. The instruction set in the main body of the program that would be required is:

movfw	**Display**	**; takes the number out of Display**
call	**_7SegDisp**	**; accesses the conversion subroutine**
movwf	**portb**	**; loads the correct code into Port B**

As you can see, nothing clever happens here. Where the actual conversion takes place is outside the main body of the program, in the subroutine. The subroutine uses the *program counter* (file register number 02). On the diagram showing the layout of file registers, this was given the name PCL – this stands for program counter latch.

The program counter

The program counter holds the address of the next instruction to be executed. There are 512 addresses in the program memory of the PIC54, so clearly the program counter must be able to hold a number as large as 511 (remember, one of the addresses is numbered 0). The PCL only holds the lower 8 bits of the program counter (bits 0 to 7). Higher bits are discussed in a later section.

Take a look at the first line of Example 2.6 (address **0043** in the program memory). While the PIC processor is executing this line, the contents of the program counter (PC) would be **0044**, as this is the next instruction to be executed. The fact the PC holds the address of the *next* instruction allows the processor to load the next instruction from the program memory *at the same time* as executing the current instruction (this is called *pipelining*). This means the processor can run through the program faster, but it is necessarily making a guess on what the next instruction will be (it guesses that it will be the next instruction in the program memory). Whenever there is a skip, a **goto**, **call**, or **retlw** this guess is incorrect, as the PC is changed. The processor then throws away (*flushes*) the instruction it guessed would be next, and loads the correct instruction. This loading takes up an extra clock cycle, and so while normal instructions take one clock cycle, skips and **goto**s, etc. take two cycles. Example 2.6 illustrates this idea – the actions of the processor during each clock cycle are provided below.

Example 2.6

```
0043  Start  btfss  portb, 0  ; tests push button
0044         goto   On
0045         goto   Off

0046  On     bsf    porta, 0  ; push button isn't pressed, so turn on LED
0047         goto   Start     ; loop back to start
      Off

0048         bcf    porta, 0  ; push button isn't pressed, so turn off LED
0049         goto   Start     ; loop back to start
```

Clock 1: The instruction at **0043** is being executed, the PC holds the number **0044** (remember, it holds the address of the *next* instruction to be executed) and in the background the processor is loading the instruction at address **0044** (**goto On**).

Clock 2: Let's say the bit being tested was clear. There is no skip, and so there was no change to the PC. The processor therefore begins executing the loaded instruction (**goto On**), increments the PC to **0045**, and in the background begins loading the instruction at address **0045** (**goto Off**). Note that it is loading the *wrong* instruction!

Clock 3: The instruction **goto On**, changes the PC to **0046**. The processor notices that the PC has changed, *flushes* the instruction it had loaded, and begins loading the instruction at address **0046** (**bsf porta, 0**). No instruction is executed during this clock cycle.

Clock 4: The processor begins executing the loaded instruction (**bsf porta, 0**), etc. . . . .

Exercise 2.14 **Challenge!** Go through the program from **Start**, this time assuming the bit being tested was set. For each clock cycle, write down the address of the instruction being executed (if any) and the value of the program counter. Do this until the program returns to **Start**. How many clock cycles does one loop take?

Now that we know what the number in the PC means, we can use this understanding to create *variable* jumps. As we have seen, skipping and **goto**, etc. are ways to change the PC; we can also change it directly as we would any other register, by acting on the PCL register (file register number 02). For example if we add the number 2 to the program counter, it will skip that many instructions:

0043	movlw	d'2'	; adds 2 to the PCL
0044	addwf	PCL, f	;
0045	goto	earth	; – not executed
0046	goto	wind	; – not executed
0047	goto	fire	; – this line is executed

While the instruction at **0044** is being performed, the PC holds **0045**. This instruction adds 2 to the PCL, changing the PC to **0047**. The processor notices the PC has changed, flushes, and begins loading the instruction at address **0047**. The code in this example is quite useless, as the skipped instructions will never be executed because the number added to the PCL is constant. However, if the number added to the PCL is variable, we can create a *look-up* table.

As you should already know, to return from a subroutine the instruction is:

> **retlw   number   ;**

However, this not only returns from a subroutine, but **ret**urns with a literal (**number**) in the working register. This instruction is key to the look-up table, and thus to the seven-segment display encoding subroutine we are trying to write:

```
_7SegDisp
        addwf   PCL, f          ; skips a certain number of instructions
        retlw   b'11101110'     ; code for 0
        retlw   b'00101000'     ; code for 1
        retlw   b'11011010'     ; code for 2
        retlw   b'11111000'     ; code for 3
        retlw   b'00111100'     ; code for 4
        etc.
        retlw   b'01010110'     ; code for F
```

Remember that this subroutine is called with the number to be displayed already in the working register. This number is then added to PCL, so the processor skips that many instructions. If the number to be displayed is 0, the processor skips 0 instructions and thus returns from the subroutine with the code to display a '0'. This applies for all the numbers being displayed (0–F).

Subroutines and the stack

We can now take a more detailed look at how subroutines work, using the program counter. When a subroutine is called, the program counter is copied into a special storage system called the *stack*. You can think of the stack as a pile of papers, so when the subroutine is called, the number in the PC is placed (*pushed*) onto the top of the stack. When a returning instruction such as **retlw** is reached, the top number on the stack is placed back in the PC (it is *popped* off the stack), thus the processor returns to execute the instruction after the original **call** instruction. In the example above we have only used one level of the stack (only one number was placed on the stack, before being taken off again). The PIC5x series have a stack which is only two levels deep (most other models have eight). When a subroutine is called within another subroutine, again the number from the PC is placed on top of the stack pushing the previous number to the level below. If you then call a third subroutine within the second, the third number goes on the top of the stack, pushing the second to the bottom level, and pushing the first number off the bottom of the stack (i.e. it is forgotten)! This means that it will not be possible to return from the first subroutine – clearly not a desirable situation. The example in Figure 2.16 illustrates this problem.

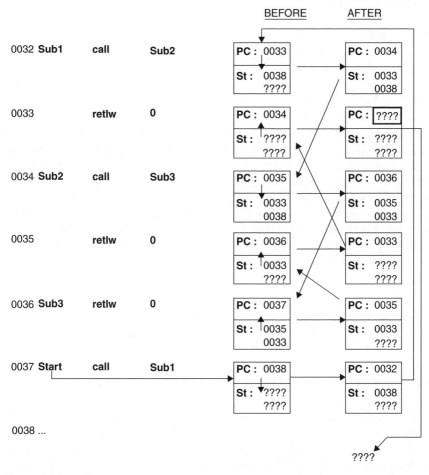

Figure 2.16

Begin where it says **Start**. When the **call   Sub1** instruction is executed, the contents of the PC are copied onto the stack. Then in the subroutine **Sub1**, when the subroutine (**Sub2**) is called, the contents of the PC are again copied onto the stack, pushing the previous value down one level. Finally, in **Sub2**, when the third subroutine (**Sub3**) is called, the PC is copied onto the stack, pushing the second down one level, and the first *out of the stack*. At the next instruction **retlw**, the number at the top of the stack is placed into the PC, thus returning from **Sub3**. Then, with the next **retlw**, the stack is again popped into the PC. However, upon the third **retlw** instruction, the processor moves an unknown number **????** from the stack into the program counter, which could make the processor effectively return anywhere (though this is *probably* the instruction at address **0000**. . . but don't count on it!). Do not worry if you find all this a bit too technical – the take-home message is: you can call a subroutine, and you can call a subroutine within a subroutine, but you cannot call a subroutine within a subroutine. Of course, this doesn't stop you calling two subroutines within the same subroutine, like this:

Sub1	**call**	**Sub2**	;
	call	**Sub3**	;
	retlw	**0**	;
Start	**call**	**Sub1**	;

One final, important word of warning: whenever you change the PCL yourself (e.g. add a number to it) *or* whenever you use a **call** instruction, bit 8 of the program counter is cleared to 0. Let's think about what this means. The 512 addresses of program memory (called a *page*) are addressed with 9 bits (bit 0–8). If bit 8 is automatically cleared, the addresses are limited to locations 0–255 (referred to as the first half of the page). The result is that all subroutines must be placed (or at least *start*) in the first half of the page, though you can call them from anywhere in the page. Furthermore, if you want to use the variable jumps described above, these too need to take place in amongst the first 256 instructions of the program.

Example 2.7

0143	**OnSub**	**bsf**	**porta, 0**	**; start of a subroutine**
0144		**retlw**	**0**	**; returns**
0145	**Start**	**call**	**OnSub**	**; tries to call sub: 'OnSub'**

While executing the **call** instruction in the example above, the PC is **0146**. The number **0146** is pushed into the stack; however, the number loaded into the program counter is *not* **0143**. Because bit 8 of the program counter is cleared by a **call** instruction, the number **0043** is placed into the PC and the processor will actually jump to address **0043** (and keep going until it reaches a return instruction).

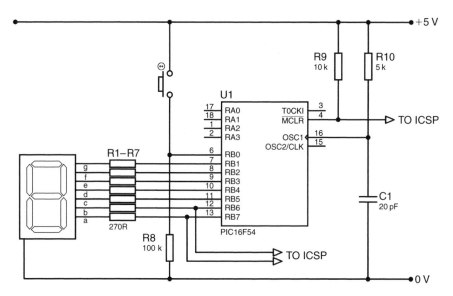

Figure 2.17

Our next project will be a counter. It will count the number of times a push button is pressed, from 0 to F. After 16 counts (when it passes F), the counter will reset to 0. The seven-segment display will be connected to pins RB1 to RB7 and the push button will go to RB0. Figure 2.17 shows the circuit diagram – pay particular attention to how the outputs to the seven-segment display are arranged. You should also note that we are using pins RB6/7 which are used for the in-circuit serial programming (ICSPCLK and ICSPDAT). If you are using ICSP, these pins should be connected *directly* to the ICSP device (such as the BFMP), as before, with a resistor between the pin and the rest of your circuit. In our case, we have resistors going between RB6/7 and the LEDs, so that's not a problem. Unfortunately, if you are powering your circuit board from the ICSP connection, you need a way to disconnect the ICSPCLK/DAT lines from your circuit when you wish to operate it, as these lines will cause some disruption. This can be achieved through a pair of DIL switches, or jumpers, which you can switch when you want to program the circuit.

The flowchart will be as shown in Figure 2.18.

The set up is much like in previous projects, but do not forget to reset any important file registers (such as the one used to hold the number of counts) in the **Init** subroutine. It may also be desirable to move the code for a **0** into Port B at the beginning (rather than simply clearing it). Testing the push button should present no problems either.

Exercise 2.15 What *two* lines will firstly test the push button, and then loop back and test it again if it isn't pressed?

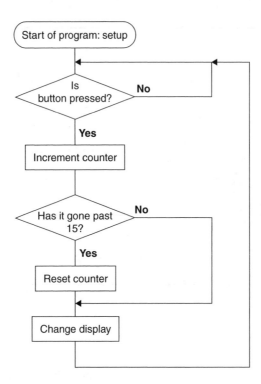

Figure 2.18

To continue developing this program it is necessary to introduce a new instruction:

> **incf   FileReg, f   ;**

This **inc**rements (adds one to) a file register, leaving the result in the file register. When the push button is pressed the program skips out of the loop. In this
case the general purpose file register which you are using to keep track of the
number of times the button has been pressed (let's call it **Counter**) should be
incremented.

Exercise 2.16 What *one* line will accomplish this?

We then need to check to see whether or not more than 15 (F in hexadecimal)
counts have been received, or in other words whether or not the number in
Counter is 16. As you know, the usual way to see whether or not the number in
a file register is a particular value is to subtract that value from the file register
(leaving the result in the working register), and then see if the result is zero. On
this occasion, however, we can simply check bit 4 of **Counter** – if low we know
it stores a number less than 16 and when it goes high we know **Counter** has
reached 16 (think about it).

Exercise 2.17 **Challenge!** What *two* lines will first test to see whether or not the number in **Counter** has reached 16, and if it has will reset **Counter** to 0 (clear it). Otherwise the program should continue, leaving **Counter** unchanged.

Finally we need to change the number in **Counter** into a seven-segment code and move it into Port B, before looping back to **Main**. This is done, as you know, using the encoding subroutine.

Exercise 2.18 Write the *four* lines that should follow the previous two, which take the number from **Counter** into the working register, call the decoding sub-routine (name it **_7SegDisp**) which returns with the correct code in the working register, and then move it into Port B. Then the program should loop back to **Main**.

Exercise 2.19 Finally, write the subroutine called **_7SegDisp** which contains the correct codes for the seven-segment display.

The program so far is shown as Program F. It is recommended that you actually build this project. Try it out and you will spot the major flaw in the project.

You should notice that when you press the button, the number 8 will appear on the display, and then when you release the button, the counter will stop on a seemingly random number between 0 and F. This is because the program isn't testing for the button to be released. So if you work out roughly how long a cycle takes in the current program when the button is pressed, you can see how often the push button is tested. There are about 11 instructions in the cycle, and we are using a 3.82 MHz oscillator. An instruction is executed once every four signals from the oscillator (at 0.96 MHz), so the cycle of 11 instructions is executed at a frequency of about 86800 Hz, that's 86800 times a second. So with the current program, if you press the button for one second, counter will count up about 86800 times (hence the 8 on the display – what you get when the display counts up through all the numbers at high speed). This project, as it is, would make a good random number generator, but let's move on.

To solve this problem we need to wait until the button is released before we test for it again. The improved program flowchart would be as shown in Figure 2.19.

All that needs to be changed is that instead of the final line **goto   Main**, we need to test the push button again. The program should go back to **Main** if it isn't pressed, and keep looping back if it is pressed.

Exercise 2.20 What *three* lines will achieve this. (**Hint:** You need to give this loop a name.)

Assemble the new program (shown in Program G), and try it out. Alas, we still have a problem.

You should notice that the counter seems to count up more than once when the push button is pressed (e.g. upon pressing the button it will go from the

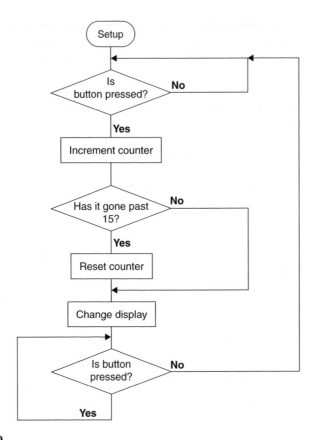

Figure 2.19

Figure 2.20

number 4 to the number 8). This jump varies in size depending on the quality of the push button used. Our problem is due to *button bounce*. The contacts of a push button actually bounce together when the push button is pressed or released. Figure 2.20 shows the signal fed to the RB0 pin.

The precise details of the bouncing vary according to button type, and indeed may be different every time the button is pressed, but button bounce is always there. As you can see from Figure 2.20 the program will count more than one signal, even though the button has only been pressed once. To avoid this, we must wait a short while after the button has been released before we test the button again. This slows down the minimum time possible between counts, but a compromise must be reached.

Example 2.8 To avoid button bounce we could wait 5 seconds after the button has been released before we test it again. This would mean that if we pressed the button 3 seconds after having pressed it before, the signal wouldn't register. This would stop any bounce, but means the minimum time between signals is excessively large.

Example 2.9 Alternatively to attempt to stop button bounce we could wait a hundred thousandth of a second after the button release before testing it again. The button bounce might well last longer than a hundred thousandth of a second so this delay would be ineffective.

A suitable comprise could be about a tenth of a second (as button bounce varies depending on the button you use, this may not be sufficient – so you may have to experiment a little). I am going to choose the longest time possible without having to use more than one postscaler. In this case the oscillator is at 3.82 MHz; divide by four to get 0.96 MHz, and then again by 256 to get the lowest frequency of the TMR0 which equals 3730 Hz. Using my own further postscaler/ marker of 255, I can get a frequency of 14.6 Hz. This total time is therefore 0.07 seconds ($=1/14.6$) which should be sufficient. The improved program flowchart is shown in Figure 2.21.

As we need to use the delay twice, we should place it in a subroutine to avoid repetition. To create a 0.07 s delay, we must wait for the TMR0 to change 255 times. At the start of the subroutine, we want to set up the *marker* register with (TMR0 + 255). Then wait for TMR0 to reach the marker, as in the previous examples. When the required time has passed, the program should return from the subroutine.

Exercise 2.21 What eight lines make up this delay subroutine?

Add the lines to call the delay subroutine at the appropriate points in the program, and the project should now work. The final program is shown in Program H.

Our next project will be a stop clock. It will show minutes (up to nine), tens of seconds, seconds, and tenths of a second, thus requiring four seven-segment displays. Using strobing, these will only require $4 + 7 = 11$ outputs. There will be a push button to start and stop the device, which will require 1 input. A second reset button can be connected to the $\overline{MCLR}$ pin without taking up an I/O pin. In this way the whole project can be squeezed onto the PIC54. RB1 to RB7 will have the seven-segment code for *all four* of the displays, RB0 will be the

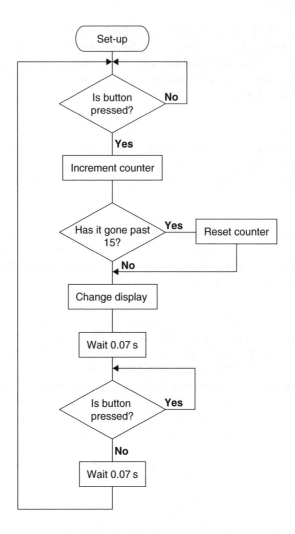

Figure 2.21

start/stop button, and finally RA0 to RA3 will control the seven-segment displays. The circuit diagram in Figure 2.22 summarizes the setup. The resistor values for the display segments are chosen in the following way. The PIC microcontroller produces a 5 V output, and the segments require 2 V and 10 mA. Therefore there is a 3 V drop across and 10 mA desired through the resistors. This would suggest a value of about $3/0.01 = 300$ ohms. However, as there are four displays being strobed, each display is only on for a quarter of the time. So to create the same brightness as if the displays were permanently on, we have to allow four times the current through, and thus quarter the resistor values. A value of 82 ohms was therefore used.

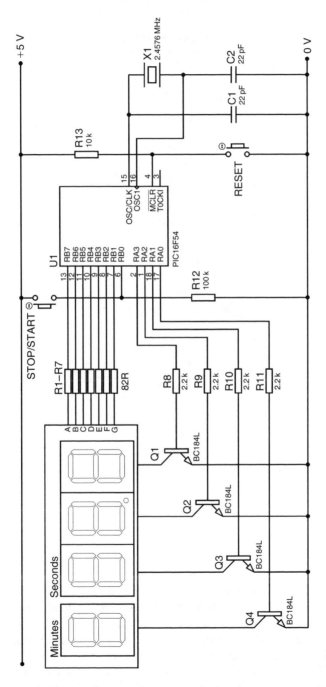

Figure 2.22

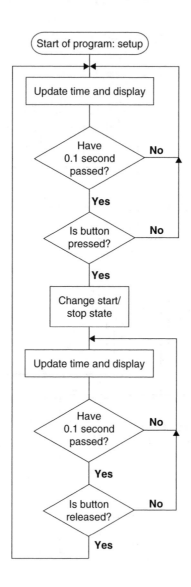

Figure 2.23

The program flowchart must now be constructed (Figure 2.23).

The flowchart is structured around testing the start/stop button, with different sections for when it has been pressed and released, while also taking care of button bounce (hence the 0.1 s delays). The box 'Update timing and displays' represents a lot of work – advancing the timing registers, keeping the displays strobing and also counting out the 0.1 s for the de-bounce routine. Tidying these linear operations into one box allows us to get a feeling for the overall structure of the program – we will put them all into a subroutine called **Update**.

We will handle the de-bouncing as follows. We will commandeer a bit in an unused GPF to act as a flag to tell us whether the button has just changed state (pressed to released, or vice versa). When this flag goes low, we should wait 0.1 s before setting it again. While the flag is low, the button will not be tested further. After the flag is re-set, we can assume the state of the button has stabilised and will continue testing its state. We will call this bit **bounce**, and assign it bit **0** of file register **08**. We can define the name of this bit, using the following command:

#define   name   FileReg, Bit

This assigns a name to a particular bit in a file register. This doesn't have to be a general purpose file register either – you can rename a bit in an SFR, such as Port A. The fundamental difference between this command and **equ**, is that a *number* must always follow **equ**, whereas *anything* can follow **#define**. The assembler will simply replace any instance of the word you have #defined with the definition you've provided.

Example 2.10

#define   LED1   porta, 0
etc.
bsf   LED1   ; turns on first LED (connected to RA0)

In the case of a general purpose bit, we naturally need to assign it to a bit in a GPF (and, of course, one which we aren't already using). I advise having one file register set aside to house any general purpose bits (you seldom need more than 8), and calling this file register **General** (or a more inspiring name if you can think of one). To define the bit **bounce** the following would be written:

#define   bounce   General, 0

If we were to write this, we would naturally have to define the file register **General**:

General   equ   08

You may than ask why we don't simply write:

#define   bounce   08, 0

The reason for this is that if I define the file register **General** as number **08**, along with all the other GPFs, there is less danger of accidentally assigning address **08** to another file register. Furthermore, people tend to feel more comfortable with names rather than numbers, so it is a good idea to use them when you can. Finally, defining of bits should take place immediately after the file register definitions in the *declaration* section of the template.

Now we have a bit which is set when the button is safe to test (more than 0.1 s have passed); this button should be set in the **Init** routine. We also require a bit to determine whether the stopwatch is in the 'start' or 'stop' state – call this bit **start** and define it as bit **1** of **General**. When this bit is *set*, we will be in the *start* state and the timer should count up. When *clear* the timing should *stop*.

The beginning of the program looks like this:

Start	call	Init	; sets up initial registers
Released			
	call	Update	; updates timing and display
	btfss	bounce	; is button safe to test?
	goto	Released	;
	btfss	portb, 0	; is button pressed?
	goto	Released	; no, so loops

In this initial loop, the program is waiting for the button to be pressed while also making sure that the timing and display is constantly up-to-date (in the **Update** subroutine). During this loop, the PIC microcontroller may be in the start *or* stop state, and so when the button is pressed, we need to *toggle* the state of the bit we called **start**. We also need to tell the program that the state of the button has just changed, so we need to activate the de-bounce routine. We will do this in a sub-routine called **PrimeBounce**. The subsequent three lines are therefore:

	movlw	b'000000010'	; toggles the state of the start bit
	xorwf	General, 1	;
	call	PrimeBounce	; activates de-bounce routine

We now enter the second loop in which the button is in the pressed state. We want to check to see if the button is safe to test (is the bit called **bounce** set?) and if so, test to see if the button has been released. Within this loop, we also need to update the timing and displays (by calling the **Update** subroutine). If the button has been released, we should activate the de-bounce routine, and then loop up to the section called **Released**.

Exercise 2.22 What *seven* lines make up this section (call it **Pressed**).

This completes the main body of the program – though clearly a lot more remains to be done in the subroutines. In the **Update** subroutine, we first test to see whether the timing routine should be active or not. Timing will take place in a subroutine called **Timer**. If **bounce** is clear, the program should be counting 0.1 s for the de-bouncing routine, which we will call **Debounce**. The following lines begin the **Update** subroutine:

Update	btfsc	start	; are we in the start or stop state?
	call	Timer	; start state, so advances timer
	btfss	bounce	; is the bounce flag low?
	call	Debounce	; yes, so calls de-bounce routine

Now all that remains in this subroutine is the handling of the seven-segment displays. This consists of two main tasks: first to choose which display it is going to turn on (tenths of second, seconds, etc.), and second, work out what to display on it. As we have a power of two as the number of displays (four is a power of two), we can use a neat trick with the TMR0 to evenly scroll through the different displays. This is the essence of strobing – first one display is turned on for a short period of time with all the others off, then it is turned off and another is turned on with its number displayed. This happens so quickly that we don't even notice it and are given the impression that all are on at the same time. We can use the two least significant bits (bits 0 and 1) of TMR0 to decide which display to turn on. If the two bits in question are **00**, tenths of second are displayed, if they are **01**, seconds are displayed, if they are **10**, tens of seconds are displayed, and finally, if they are **11**, then minutes are displayed. How do we just look at the two least significant bits? How do we ignore the rest of the number? The answer is *ANDing*. The logic command AND takes a certain number of bits as its inputs (in the case of a PIC program it takes *two*) and depending on their states creates an output (i.e. the result of the logic operation). The table below (known as a *truth table*) shows the effect of the AND command on two bits.

Inputs		Result
0	0	0
0	1	0
1	0	0
1	1	1

As you can see, the output bit is high if the first **and** second input bits are also high. A useful property of this command is that if you AND a bit with a **0** the bit is *ignored*, and if you AND a bit with a **1**, the bit is *retained*.

Example 2.11 ANDing the two 8 bit numbers **01100111** and **11110000**, produces the following result:

$$01100111$$
$$11110000$$
$$\overline{}$$
$$01100000$$

Notice how by ANDing the top number with **11110000**, bits 4 to 7 are retained (kept the same), whereas bits 3 to 0 have been ignored (replaced with 0). In this way we can ignore bits 2 to 7 of TMR0, retaining only bits 0 and 1.

Exercise 2.23 What number must TMR0 be ANDed with to ignore all but bits 0 and 1?

The instruction that allows us to AND two numbers together is:

andlw   number   ;

This **AND**s the literal (**number**) with the number in the working register. However an alternative instruction more suited to this example is:

andwf   FileReg, f   ;

This **AND**s the number in the working register with the number in a file register, leaving the result in the file register. It would be quite disastrous to actually affect the number in TMR0 as this would mess up the whole of the timing side of things, so we replace the **,f** with a **,w**, so that the result is placed in the working register, leaving the file register unchanged. The instruction pair used to ignore all but bits 0 and 1 of TMR0, leaving the result in the working register as:

movlw   b'00000011'   ; ignores all but bits 0 and 1 of TMR0
andwf   TMR0, w   ; leaving the result in the working
**   ; register**

How do we use this number to select which display we turn on? We simply add the result (a number between 0 and 3) to the program counter, and have several jumping (**goto**) instructions afterwards which are executed depending on the result:

addwf   PCL   ; adds the result to the program
**   ; counter**
goto   Display10th   ; displays tenths of a second
goto   Display1   ; displays seconds
goto   Display10   ; displays tens of seconds
goto   DisplayMin   ; displays minutes

The program thus branches out to different sections depending on the two least significant bits of the TMR0. These sections will take the following form:

Display10th   movfw   TenthSec   ; takes the number out of TenthSec
**   call   _7SegDisp   ; converts the number into 7-seg**
**   ; code**
**   movwf   portb   ; displays the value through Port B**

**   movlw   b'0010'   ; turns on correct display**
**   movwf   porta   ;**

**   retlw   0   ; returns**

You may have noticed that for a brief time, the *wrong* number is being displayed on a display; this is of no consequence as it is on the wrong display for about

300 000th of a second. If you are a perfectionist, or find in other cases that there is a considerable delay between putting the correct number in Port B, and turning on the correct display, simply clear Port A before changing the number in Port B. No display is better than a wrong display (for a short period of time). The other sections will be like this, except with a different file register used as the source of the number being displayed, and a different number being moved into Port A.

Exercise 2.24 Write the other three sections required to finish the display manager and therefore completing the **Update** subroutine.

The **Timer** subroutine is not simply a delay as we've used before – it should check whether a certain amount of time has passed, and if it hasn't, it should return to allow the program to continue with other tasks. This subroutine will first have to tell whether or not a tenth of a second has passed, as this is the smallest unit of time being displayed. The TMR0, when prescaled by the maximum amount of 256, counts up 2400 times a second, and thus 240 times in a tenth of a second. We can therefore time this using just one marker, which we will call **Mark240**. The first part of the timing subroutine will be reasonably similar to the delay instruction set, but with return instructions where previously there were looping instructions:

```
Timer   movfw   Mark240      ; test to see if TMR0 has passed
        subwf   TMR0, w      ; 240 cycles (i.e. 1/10th of a second
        btfss   STATUS, Z    ;   has passed)
        retlw   0            ; hasn't passed, so returns

        movlw   d'240'       ; has passed – resets marker
        addwf   Mark240, f   ;

        incf    TenthSec, f  ; increments number of tenths of a
                             ;   second
```

Rather than looping back to **Timer** if the correct time hasn't elapsed, the program returns from the subroutine, enabling it to go on and perform the other necessary tasks. Also note that the number 240 must have been moved into **Mark240** to begin with (e.g. in the **Init** subroutine). As shown above, once a tenth of a second has passed, the file register **TenthSec** is incremented (one is added to it). In this way the file register **TenthSec** holds the number of tenths of a second which have passed, and thus can be used easily in the display section. (If **TenthSec** counted down from 10 to 0, for example, it *wouldn't* hold the actual number of tenths of second which had passed.) Once a tenth of a second has passed, we need to check whether a whole second has passed (i.e. if 10 tenths of a second have passed). So we use the technique always when checking whether a file register has reached a certain number – we subtract that number from the file register, leaving the result in the working register, and then test to see whether or not the result is zero:

```
movlw    d'10'            ; tests to see whether TenthSec has
subwf    TenthSec, w      ; reached 10 (i.e. whether or not one
                          ; second has passed)
btfss    STATUS, Z        ;
retlw    0                ; 1 second hasn't passed, so returns

clrf     TenthSec         ; 1 second has passed, so resets
incf     Seconds, f       ; TenthSec and increments the
                          ; number of seconds
```

This instruction set is much the same as the one for tenths of a second, except the number we are testing for will always be 10, and we reset back to 0 when the correct time has elapsed. Further sections for tens of seconds and minutes will take much the same form as the one above.

Exercise 2.25 Write the instruction sets to continue the timing subroutine from the line **incf Seconds, f**, for tens of seconds, and then for minutes. (**Hint:** The last line should be **incf Minutes, f**.)

The next step is to test to see if **Minutes** has reached 10. At this point the stop clock's maximum is reached, and device should reset – all that is required is clearing **Minutes**, as all the other file register will have reset 'on the way'.

```
movlw    d'10'            ; test to see whether Minutes has
subwf    Minutes, w       ; reached 10
btfss    STATUS, Z        ;
retlw    0                ; 10 minutes haven't passed, so returns

clrf     Minutes          ; 10 minutes have passed, so resets
retlw    0                ; Minutes and returns
```

This completes the **Timer** subroutine. Make sure you set up the timing registers with appropriate values in the **Init** routine. This only leaves two subroutines associated with de-bouncing: **PrimeBounce** and **Debounce**. The **Debounce** subroutine is run if and only if the **bounce** flag is cleared. It should determine whether or not roughly 0.1 second has passed, and if so, it should set the **bounce** flag. I've used a marker of 250 to count for just over 0.1 second:

Debounce
```
movfw    Mark250          ; if about 0.1 second has
subwf    TMR0, w          ; passed, sets the bounce
btfss    STATUS, Z        ; bit
retlw    0                ;
bsf      bounce           ;
retlw    0
```

Therefore, in **PrimeBounce**, the **bounce** flag needs to be cleared to activate the **Debounce** routine. The marker **Mark250** also needs to be initialised with the value of TMR0 + 250:

PrimeBounce

```
       bcf      bounce        ; clears bounce bit to trigger
       movlw    d'250'        ;  and sets up Mark250 so that
       addwf    TMR0, w       ;  about 0.1 second will be
       movwf    Mark250       ;  counted
       retlw    0             ;
```

The entire program (it's quite a large one!) is now complete and is shown in its entirety in Program I. You will, I hope, find the end result much more satisfying than previous examples, but will recognise a lot more work went into it. When constructing a program of that size (or larger) I cannot stress enough the importance of taking breaks. Even when it is really flowing and you are really getting into your program, if you step back for ten minutes and relax, you will return looking at the big picture, and may find you are overlooking something simple. Good planning with flowcharts and diagrams will help prevent such oversights significantly. You should also talk to people about decisions you should make along the way – even if they may not know the answer any more than you do, simply asking the question and talking it through helps you get it straight, and the majority of the time you will end up answering your own question.

Logic gates

After a long and complicated project, let's return to something simpler. You've already seen three logic gates (inclusive OR, exclusive OR, and AND), and we'll now look at the other five (NOR, NAND, BUFFER, NOT and XNOR). The truth tables for the new gates are as follows:

NOR

Inputs		Result
0	0	1
0	1	0
1	0	0
1	1	0

The result is the opposite of an inclusive OR gate (i.e. **not** an inclusive OR gate).

NAND

Inputs		Result
0	0	1
0	1	1
1	0	1
1	1	0

The result is the opposite of an AND gate (i.e. **not** an AND gate).

BUFFER

Input	Result
0	0
1	1

Only one input is used, the output copies the input.

NOT

Input	Result
0	1
1	0

Again only one input, but the output is the opposite of the input (i.e. **not** the input).

XNOR

Inputs		Result
0	0	1
0	1	0
1	0	0
1	1	1

The result is the opposite of an exclusive OR gate (i.e. **not** an exclusive OR gate).

There aren't instructions for all these gates, but all can be constructed through a combination of those which we are given. The project to experiment with the use of these gates and their instructions will be a multi-gate IC (a chip which will effectively act as any of these eight gates). There will be two inputs and one output which are the actual parts of the artificial gate. There will also be three bits for choosing the type of gate being simulated. Three bits can select a total of eight variations (000 to 111). There will be one combination for each of the eight logic gates. These selection bits will be RA1 to RA3, and the inputs of the gate will be RB0 (main input) and RA0 (secondary input), with the gate output at RB4. The circuit diagram is shown in Figure 2.24.

The flowchart must now be constructed.

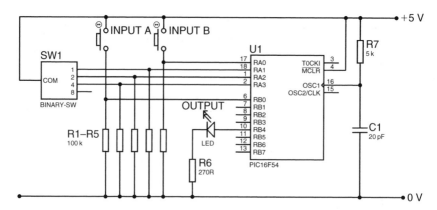

Figure 2.24

Exercise 2.26 Have a go yourself at constructing the flowchart, before looking at my version in the answer section (Appendix I). Remember, as long as the gist of it is the same, it isn't crucial that the minor details are the same as mine, but you need not make it more than three boxes in size, as we aren't yet concerned with sorting out how to manage the imitating of the individual gate types.

The encoding we will use is shown in Table 2.2.

Table 2.2

RA3-RA1	Logic Gate
000	Buffer
001	AND
010	IOR
011	XOR
100	NOT
101	NAND
110	NOR
111	XNOR

In this way, RA1 and RA2 determine the principle type of gate, and RA3 determines whether or not the result should be inverted. We wish to use the branching technique of adding a number to the program counter, which was discussed earlier. Unfortunately the number we want to add is found in bits 1 and 2 of Port A. We could simply ignore the other bits (using ANDing), but this would leave us with a number which could go up to 6 (0000 to 0110). What we really want to do is *rotate* the number to the right (e.g. making 0110 into 0011).

rrf   FileReg, f   ;

This rotates the number in a file register to the right, leaving the result in the file register. Its complementary instruction is:

rlf   FileReg, f   ;

This rotates the number in a file register to the left, again leaving the result in the file register. You may wonder where the bit that gets 'bumped off' goes, and where the bit that fills the gap left comes from. There is an intermediate bit called the *carry flag*. This is one of the flags (like the zero flag) in the STATUS register. It has other purposes as well as that shown in Figure 2.25.

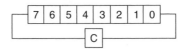

Figure 2.25

So when rotating right, the state of bit 0 is moved into the carry flag, and the previous state of the carry flag is moved into bit 7. This is a consequence of the carry flag's main property which will be discussed at a later stage. It is important to clear the carry flag before any rotation instruction, because otherwise, if set, it will put a one where a gap was left upon rotation – in most cases this is undesirable. To thus be able to use the number in Port A to branch to the correct place, the following is done to it:

```
Main    bcf     STATUS, C    ; makes sure carry flag is clear
        rrf     porta, w     ; bumps off bit 0, leaving the result in
                             ;   the working register
```

Bits 2 and 3 of the working register should then be masked (leaving a result that is between 00 and 11) using the **andlw** instruction. The result is then added to the PC to branch to the correct section:

```
        andlw   b'0011'      ; masks bits 2 and 3
        addwf   PCL, f       ; branches to correct gate section
        goto    BufferNOT    ; handles Buffer and NOT gates
        goto    ANDNAND      ; handles AND and NAND gates
        goto    IORNOR       ; handles IOR and NOR gates
```

XORXNOR
```
        etc.
```

We don't need to add a fourth **goto** command as the XOR/XNOR section can simply follow on from the above. In this section, we take the number from Port A and XOR it with Port B (in doing this, bit 0 of each will be XORed). We then test RA3 – if it's set, the PIC microcontroller is emulating the negative equivalent gate (i.e. XNOR, rather than XOR), so this bit should be inverted. Bit 0 of the result is the output that we wish to move into RB4. This could be done using testing instructions (**btfss**) and setting/clearing instructions (**bsf** and **bcf**), but a more cunning method employs the following command:

```
        swapf   FileReg, f   ;
```

This **swaps** the lower nibble (bits 0 to 3) with the upper nibble (bits 4 to 7) of a file register, and leaves the result in the file register.

Example 2.12

```
        movlw   b'00110101'  ; moves a number into ABC
        movwf   ABC          ;
        swapf   ABC, f       ;
```

The number in the file register ABC is now **b'01010011'**.

Exercise 2.27 What would be the resulting number if the following number was 'swapped': **b'00000001'**?

Thus if we swap the file register holding the result of the XOR operation, the state of bit 0 will be swapped into bit 4. The code for the XOR/XNOR section is shown below:

XORXNOR

```
        movfw   porta       ; reads Input B
        xorwf   portb, w    ; XORs with Input A
        movwf   STORE       ; stores result in STORE
        btfsc   porta, 3    ; tests RA3
        comf    STORE       ; inverts answer, if necessary
        swapf   STORE, w    ; swaps nibbles (bit 0 → bit 4)
        movwf   portb       ; outputs result
        goto    Main        ; loops back to start
```

Note that we keep the result of the **xorwf** operation in the working register, rather than in Port B. This is because any bits configured as inputs would essentially ignore the result of the XOR operation, and remain at the value dictated by the circuit outside the PIC microcontroller. Only bits configured as outputs would actually change. The result is kept temporarily in a GPF we've called **STORE**, inverted if RA3 was high, then swapped so that the result bit is held in bit 4.

The AND/NAND section is identical to the XOR/XNOR section above, with the exception of one line (replace **xorwf** with **andwf**). Rather than copy out the section again and waste program memory space, We can give the line after the **xorwf** instruction a label: **Common**. The AND/NAND section is therefore:

ANDNAND

```
        movfw   porta       ; reads Input B
        andwf   portb, w    ; ANDs with Input A
        goto    Common      ; rest is same as XOR/XNOR
```

Exercise 2.28 What *three* lines make up the IOR/NOR section, and what *two* lines make up the Buffer/NOT section.

The program is now complete and the whole lot is shown in Program J.

The watchdog timer

One of the useful properties of the PIC microcontroller is its *watchdog timer* – an on board timer which is driven by a resistor/capacitor network which is actually inside the microcontroller. It is thus completely independent of external components. The watchdog timer steadily counts up, and when it reaches its maximum, the PIC microcontroller will automatically reset. It is thus quite useful in devices where it is not a great problem to be constantly resetting (for at least most of the

Table 2.3

PS2	PS1	PS0	Prescaling rate
0	0	0	1:1
0	0	1	1:2
0	1	0	1:4
0	1	1	1:8
1	0	0	1:16
1	0	1	1:32
1	1	0	1:64
1	1	1	1:128

time), e.g. alarm systems. It is used as a safety feature such that if the program crashes for some reason, the program will soon reset and resume normal operation. The time for the watchdog timer to cause a *timeout* (for it to cause a reset) varies from 18 milliseconds to 2.3 seconds depending on the amount of prescaling. You can prescale it using the OPTION register (you may remember this from when we studied the TMR0). If left unprescaled it will cause a timeout after 18 ms. To prescale it, bit 3 of the option register must be set, thereupon bits 2 to 0 decide how much it is prescaled by (Table 2.3).

The maximum prescaling (128) will cause it to timeout after (0.018×128) seconds = 2.304 seconds. There is, however, no way to simply turn the watchdog timer off unless you don't need it at all (in which case you disable it using the configuration bits when writing your program to the PIC microcontroller). If it is needed for part of the program, how do you stop it from causing resets during the rest of the program? The answer is constantly resetting it. The instruction for this is:

clrwdt   ;

This **clears** the **watchdog timer** (i.e. makes it 0), and thus resets it. This must be done at specific intervals to stop the watchdog timer reaching its maximum and thus causing the timeout, i.e. if the watchdog timer resets the PIC microcontroller after 18 ms, then you need the **clrwdt** instruction to be executed at least once every 18 ms.

To try out the watchdog timer, the next project will be an alarm system. There will be a signal coming from a motion sensor at RA0 (it can be simulated by a push button), and a siren (or buzzer) at RA3 to indicate when the alarm has been set off. A toggle switch (RA1) will either set, or disable the alarm, a green LED (RB0) will show the alarm is disabled, and a red LED (RB1) will show it to be set. To conserve battery life the LEDs will flash rather than stay turned on, flashing on for one tenth of a second every 2.3 seconds (this number should sound familiar). Once triggered, the siren will go on indefinitely until the device is turned off. You may want to make an addition whereby it turns off after 20 minutes, but this is not investigated in this example. The circuit diagram is shown in Figure 2.26.

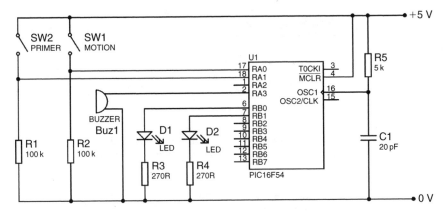

Figure 2.26

The flowchart for the program is as shown in Figure 2.27.

As in the previous example, you are now expected to write most of the program, but naturally you will be guided through each step. To begin with, as we *are* using the Watchdog timer in this program, we should change the configuration bits accordingly. The configuration line at the top of the program should therefore be:

__config   _RC_OSC & _WDT_ON & _CP_OFF

Exercise 2.29 Write the *three* lines used to first test the setting switch, and thereupon jump to another part of the program called **GreenLed** if the bit is high or simply turn on the red LED if the bit is low (as shown in Figure 2.27).

Exercise 2.30 Write the *two* lines which make up the section called **GreenLed**, in which the green LED is turned on, and then the program jumps back to a section labelled **TenthSecond**.

Exercise 2.31 Write the *seven* lines which will make up the section which tests to see whether or not a tenth of a second has passed, and turns off all LEDs if such a time has passed. In either case it then moves on to the rest of the program. Don't forget that as you are using the prescaler for the WDT, TMR0 is not prescaled. You will therefore need to slow it down by 256 (a task which is normally done by the prescaler) yourself. This is best done by moving TMR0 into the working register, and then testing the zero flag. If it is set, the TMR0 has reached zero, and you may continue on to the next postscaler (after incrementing TMR0), otherwise skip everything by going to the next section labelled **Continue**. This next postscaler should be around 240 because 2400/10 = 240, but could vary depending on what is easiest. Do not forget to reset the postscaler after it has reached 0; however the correct number should be moved into it to start with in the **Init** subroutine. Having said this, if your postscaler is 256, you don't need to reset it . . . think about it.

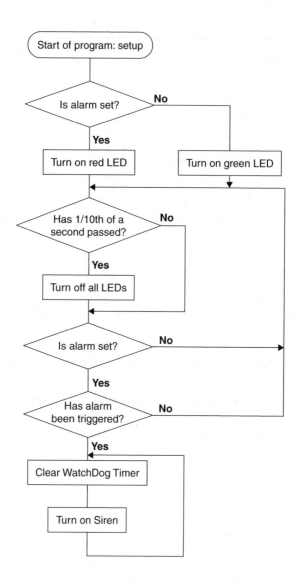

Figure 2.27

The next step is to test the setting button once more, to see whether or not it should react to the alarm being triggered. If the alarm is disabled (the bit is high), the program should return to the **TenthSecond** section, otherwise it should continue.

Exercise 2.32 What *two* lines will achieve this?

If the program continues, the alarm is set, and the trigger bit (RA0) should be tested. If no signal is received, the program should loop back to **TenthSecond**, or otherwise continue.

Exercise 2.33 What *two* lines will achieve this?

If the motion sensor has been set off the siren should be turned on, and the program should enter a cycle where the watchdog timer is constantly being reset.

Exercise 2.34 What *three* lines will finish the program?

Final instructions

There are only four more instructions which you haven't yet come across. You should be able to guess the functions of the first two of these – **decf** and **incfsz** – as they are just like their counterparts.

    decf   FileReg, f   ;

This **dec**rements (subtracts one from) the number in a file register, leaving the result in the file register.

    incfsz   FileReg, f   ;

This **inc**rements (adds one to) the number in a file register leaving the result in the file register. If this result is zero the program will skip the next instruction.

The next instruction may seem absolutely pointless but *does* actually come in quite handy every now and then:

    nop   ;

This stands for **n**o **op**eration, and does nothing.

Finally, if you are tired by now, you'll be pleased to learn the last instruction to be learnt is:

    sleep   ;

As you may have guessed, this sends the PIC microcontroller to sleep (a special low power mode). The outputs will stay the same when the PIC microcontroller goes into sleep, and can be woken up by a watchdog timer timeout, or an external reset (from the $\overline{\text{MCLR}}$ pin). A useful application combining both the **sleep** instruction and the watchdog timer allows devices to appear to automatically turn on. If, for example, a device were to turn on when moved, the program should test a vibration switch, go to sleep (until reset by the watchdog timer) if there is no movement, or alternatively skip out of the loop and constantly reset the watchdog timer as it carries on through the rest of the program, if there is movement. In this way the PIC microcontroller would be in a low power consuming mode for most of the time (it is effectively off), and would come to life when movement is detected. Figure 2.28 demonstrates this best.

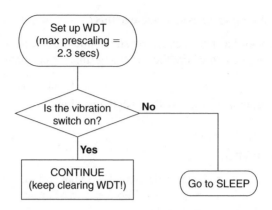

Figure 2.28

The STATUS file register

Just before we move on to the final program in this chapter, we will examine the STATUS file register in greater detail.

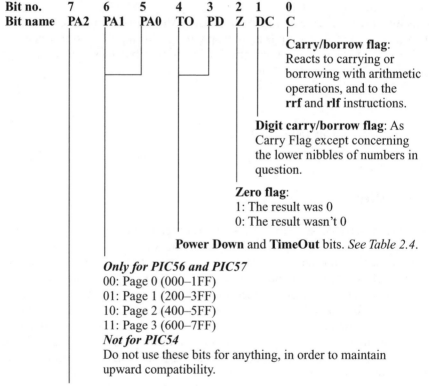

Bit no.	7	6	5	4	3	2	1	0
Bit name	PA2	PA1	PA0	TO	PD	Z	DC	C

Carry/borrow flag: Reacts to carrying or borrowing with arithmetic operations, and to the **rrf** and **rlf** instructions.

Digit carry/borrow flag: As Carry Flag except concerning the lower nibbles of numbers in question.

Zero flag:
1: The result was 0
0: The result wasn't 0

Power Down and **TimeOut** bits. *See Table 2.4.*

Only for PIC56 and PIC57
00: Page 0 (000–1FF)
01: Page 1 (200–3FF)
10: Page 2 (400–5FF)
11: Page 3 (600–7FF)
Not for PIC54
Do not use these bits for anything, in order to maintain upward compatibility.

Not for PIC5x series
Do not use this bit for anything, in order to maintain upward compatibility.

Table 2.4 Power down and timeout bits

TO	PD	Reset caused by ...
0	0	WDT wakeup from sleep
0	1	WDT timeout (not during sleep)
1	0	MCLR wakeup from sleep
1	1	Power-up

There are three new concepts introduced: the *digit carry flag*, the business of *pages* of memory, and the two bits which we can test to find the reason behind the PIC microcontroller resetting.

The carry and digit carry flags

The digit carry flag is affected only by addition and subtraction instructions. Think of the numbers in question (being added or subtracted) in hexadecimal.

```
    C   DC
        X X
 +      X X
        X X
```

The digit carry flag is set if something is carried over when adding the lower nibbles of two numbers together, and clear if nothing is carried.

Example 2.13 When adding 56h and 3Ah, we first add the lower nibbles: A and 6. These add together to make 16, or in other words, leave 0 and carry a 1. Because a one *is* being carried, the digit carry flag is *set*. We now add 5, 3, and 1 (carried over) making 9. *Nothing* is carried over so the carry flag remains *low*.

```
    0   1
        5 6
 +      3 A
        9 0
```

Example 2.14 When adding 32h and F5h, we first add the lower nibbles: 2 and 5. These add together to make 7, or in others words, leave 7 and carry nothing. Because *nothing* is being carried, the digit carry flag is *clear*. We now add 3 and F making 18, or in other words 2 and carry a 1. Because a one *is* being carried, the carry flag is *set*.

```
    1   0
        3 2
 +      F 5
        2 7
```

When subtracting, both act as $\overline{\text{borrow}}$ bits, i.e. if something *is* borrowed when subtracting, they are *clear* and vice versa. (The bar over the name, as with the $\overline{\text{MCLR}}$, means that it is active low – triggered by a negative result.) The digit carry ($\overline{\text{borrow}}$) flag again concerns the lower nibbles, and the carry ($\overline{\text{borrow}}$) flag the upper nibbles.

$$
\begin{array}{r}
\text{cX} \quad \text{dcX} \\
-\quad \text{X} \quad\ \ \text{X} \\
\hline
\text{X} \quad\ \ \text{X}
\end{array}
$$

Example 2.15 When subtracting 6Bh from 8Dh, we first subtract the lower nibbles (B from D). These leave 2, borrowing nothing. Because *nothing* is borrowed, the digit carry ($\overline{\text{borrow}}$) flag is *set.* We now subtract 6 from 8, leaving 2 and borrowing nothing. The carry ($\overline{\text{borrow}}$) flag is therefore also set.

$$
\begin{array}{r}
^0 8 \quad ^0 D \\
-\quad 6 \quad\ \ B \\
\hline
2 \quad\ \ 2
\end{array}
$$

Example 2.16 When subtracting 7Eh from 42h, we first subtract the lower nibbles (E from 2). We need to borrow 1, making the subtraction 12h − E. This leaves 4, borrowing 1. Because one *is* borrowed, the digit carry ($\overline{\text{borrow}}$) flag is *clear*. We now subtract 7 from 3 (4 − 1 which was borrowed). We again need to borrow 1, making the subtraction 13h − 7. This leaves C, borrowing 1. Because one *is* borrowed, the carry ($\overline{\text{borrow}}$) flag is therefore also *clear*.

$$
\begin{array}{r}
1(4-1) \quad 12 \\
-\quad 7 \qquad\ \ E \\
\hline
C \qquad\ \ 4
\end{array}
$$

This result is effectively a negative number (C4 in this case corresponds to −3C).

To summarise the effect of subtraction on the carry flag: if the result is negative it is clear, and if it is positive (or zero) it is set. The same applies to the digit carry flag except that you look at the lower nibbles, rather than the whole number when performing the subtraction.

Pages

We turn now to this business of pages. You may remember that the PIC54 has 1FFh bytes of program memory (up to 512 instructions). Other members of the PIC5x series can have more than this: the PIC57 has 7FFh bytes (up to 2048

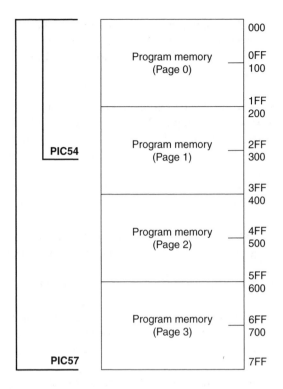

Figure 2.29

instructions, or '2k'). From this we can deduce that while the program counter in the PIC54 is 9 bits long, it is 11 bits long in the PIC57! The program memory is distributed into pages (blocks of program memory of size 512 instructions), as shown in Figure 2.29.

As discussed on page 50, bit 8 of the PC chooses the first or second half of the page. For multi-page PIC models (such as the PIC57), there are two extra bits (bits 9 and 10) in the program counter which select which page is active. These bits are mapped in the STATUS register – bits 5 and 6 (called PA0 and PA1). Let's look at pages in three situations: 'sequential operation' (running through the program in order), **goto**, and subroutines.

1. Sequential operation In this mode, you can ignore the PC. It counts up and crosses page boundaries without you having to worry about it. However, bits PA0 and PA1 of STATUS will remain unchanged, even if you move into a higher page – they are simply a way to *force* the PC to change page. In summary, the PIC processor will happily step through the instructions in the program shown below.

00FF		bsf	porta, 0	; turns on LED
0100		btfss	portb, 0	; tests button
0101		bcf	porta, 0	; turns off LED
0102		etc.		

2. Goto With the **goto** we have a problem, in that we can only specify bits 0 to 8 of the address we want to jump to. This means we can only jump to addresses which are in the same page in the program memory, i.e. if we are in Page 0, we cannot use **goto** alone to jump to a location in Page 1. What we can do is set bits PA0 and PA1 according to the page we wish to jump to. The PC will ignore these bits *until it comes to a **goto** (or **call**) instruction.* When it reaches a **goto** instruction, it will jump to the address specified by the **goto** and PA0/1 bits.

0043		bsf	porta, 0	; turns on LED
0044		goto	Wind	; branches to section in Page 1
etc . . .				
0254	Wind	bsf	porta, 1	; activates windmill

In the above example, when the processor reaches the instruction at address **0044**, it *won't* jump to the **Wind** section. Instead, it will jump to address **0054** (i.e. the equivalent of **0254** of Page 0). The correct approach would be:

0043		bsf	porta, 0	; turns on LED
0044		bsf	STATUS, PA0	; selects Page 1
0045		bcf	STATUS, PA1	; selects Page 1
0046		goto	Wind	; branches to section in Page 1
etc . . .				
0254	Wind	bsf	porta, 1	; activates windmill

Note that although I only needed to set PA0 in order to select Page 1 (from Page 0), I have also cleared PA1 just to make sure.

Exercise 2.35 What *three* lines are needed to jump from a location in Page 1 of the program, to a location Page 3 (labelled **Earth**).

3. Subroutines First, you should remember that bit 8 of the PC is cleared upon any **call** instruction, or when the PCL is changed by the user (e.g. a number is added to it). As discussed earlier, this means that all subroutines have to take place in the top half of any given page. The *stack*, which stores the address in the program memory after the **call** instruction was made (the address which the processor should return to after executing the subroutine), is as wide as the PC. This means the stack will always correctly remember the point in the program where the subroutine was originally called. However, just like the **goto**, the **call** instruction on its own cannot specify a location outside the current page. To call a subroutine in another page (remember – it must be in the top half), set the

STATUS bits (PA0 and PA1) before the **call** instruction:

0043	Roast	btfss	portb, 4	; checks temperature
0044		retlw	0	; too cold
0045		retlw	1	; too hot
etc.				
04E2		bcf	STATUS, PA0	; selects Page 0
04E3		bcf	STATUS, PA1	; selects Page 0
04E4		call	Roast	; calls subroutine in Page 0
04E5		etc.		

The subroutine **Roast** is called, and the number **04E5** is placed onto the stack. Upon reaching the **retlw** command, **04E5** is loaded back into the program counter, and the processor continues where it left off, in Page 2.

What caused the reset?

The PowerDown and TimeOut bits can be read at the beginning of the program to see what made the PIC microcontroller reset (i.e. why is it at the start of the program). This could simply be due to the fact that it had just been turned on (power-up), or alternatively due to WDT timeout. This may be important because you may not want the program to do the same thing (e.g. setting up, or perhaps clearing, of file registers) when it first starts up, as when it is reset by the WDT for example (Table 2.5).

Example 2.17 To make the program call the **Init** subroutine when the PIC microcontroller is first powered up, but not when reset for any other reason (i.e. just skip the **call Init** line), the following instruction set is used:

Start		btfsc	STATUS, 3	; tests PowerDown bit
		btfss	STATUS, 4	; PD is 1, test TimeOut bit
		goto	Main	; PD is 0, or TO is 0, so skips Init
		call	Init	; PD and TO are 1, so calls Init
Main		etc.		

Exercise 2.36 Make the program test to see whether there was a WDT timeout, or see if it's just powered up. If it has just powered up call a subroutine called PreInit, otherwise carry on.

Table 2.5

TO(4)	PD(3)	Reset caused by ...
0	0	WDT wakeup from sleep
0	1	WDT timeout (not during sleep)
1	0	MCLR wakeup from sleep
1	1	Power-up

Indirect addressing

There remains one more concept – that of *indirect addressing*. You may have noticed two file registers (*indirect address* (**00**) and *FSR* (**04**)) have not been explained yet, and these are both involved in this concept. This is probably the hardest idea to fully grasp and so it will be explained twice. First I will introduce it technically, then give an analogy which should make it easier to understand.

Think of storing a number (**N**) in a general purpose file register; you would move the number **N** into (for example) file register number **09**. This is *direct addressing*. However, you could also tell the program to move the number **N** into file register number **X**, where the file register called **X** holds the number **09**. This is called *indirect addressing*. The file register **X** is actually called the **file select register** (because it is a file **register** which **selects** which **file** register to move a number into). To use indirect addressing, move the number you wish to be stored into the **indirect address**. The indirect address is therefore not a file register as such, merely a gateway to another file register.

If you are still confused by this stage (I don't blame you), the following analogy should set things straight. Think of the **indirect address** as a envelope, and the **file select register** as the address on the envelope. When you use indirect addressing you put the number in an envelope, and it is sent to the address on the envelope (just as with our own reliable post service except with a delivery time of roughly 0.000001 second it is slightly faster!).

Example 2.18 Move the number **00** into file registers numbers **08** to **1F**.

Rather than writing:

```
clrf     08      ; clears file register number 08 (it hasn't
                 ;   been given a name)
clrf     09      ; clears file register number 09
clrf     0A      ; clears file register number 0A
etc. . . .
clrf     1F      ; clears file register number 1F
```

. . . we can use indirect addressing to complete the job in fewer lines. The first address we want to affect is **08**, so we should move **08** into the **file select register** (the address on the envelope):

```
movlw    d'08'   ; moves the number 08 into the FSR
movwf    FSR     ;
```

We then send the number **00** through the 'post' by moving it into the **indirect address** (the envelope). The instruction **clrf** effectively moves the number **00** into the file register (thus clearing it):

```
clrf     INDF    ; clears the indirect address
```

File register **08** has now been cleared (whatever you now do to the **INDF** you actually do to file register number **08**). We now want to clear register **09**, we thus increment the **FSR** (add one to it), so now whatever you do to the **INDF** you actually do to file register number **09**.

```
incf    FSR              ; increments the FSR
```

The program can now loop back to the line where the **INDF** is cleared. However it must first check to see whether or not the FSR has passed the file register **1F**, in which case it should jump out of the loop. To see whether a file register holds a particular number, you subtract that number from the file register and see whether or not the result is zero:

```
movlw   20h        ; has the FSR reached the hexadecimal
subwf   FSR, w     ;  number 20?
btfss   STATUS, Z  ;
goto    ClearLoop  ; it hasn't, so keep looping
                   ; it has, so exits loop
```

The following instruction set is very useful to put in the **Init** subroutine to systematically clear a large number of file registers:

```
            movlw   d'08'      ; moves the number 08 into the FSR
            movwf   FSR        ;
ClearLoop   clrf    INDF       ; clears the indirect address
            incf    FSR        ; increments the FSR
            movlw   20h        ; has the FSR reached the
            subwf   FSR, w     ;  hexadecimal number 20?
            btfss   STATUS, Z  ;
            goto    ClearLoop  ; it hasn't, so keep looping
                               ; it has, so exits loop
```

You can adjust the starting and finishing file registers (at the moment **08** and **1F** respectively).

The **FSR** has a secondary purpose, as well as indirect addressing. It is used to select GPFs on larger PIC microcontrollers such as the PIC57. As well as general purpose file registers at addresses 08-1Fh, this particular PIC microcontroller also has available space at addresses 30-3Fh, 50-5Fh, and 70-7Fh (that's 48 extra file registers!). However, these extra addresses cannot be accessed in the same way as the others. Bits 5 and 6 of the **FSR** are used to select which set of registers we wish to access (read or write to), as shown in the Table 2.6.

For example, let's say I have made the following declarations:

```
Tailor    equ    15h
Tinker    equ    35h
Soldier   equ    55h
Spy       equ    75h
```

Table 2.6

FSR		File registers selected
Bit 6	Bit 5	
0	0	10-1F
0	1	30-3F
1	0	50-5F
1	1	70-7F

Note that file registers 00-0F are independent of the FSR and can be accessed regardless of the state of these two bits.

If I want to write a number to the file register **Tinker**, I need to do the following:

bsf	**FSR, 5**	**; selects file registers 30-3F**
bcf	**FSR, 6**	**; selects file registers 30-3F**
movlw	**d'30'**	**; moves number into Tinker**
movwf	**Tinker**	**;**

Note that without the first two lines setting the correct bits in **FSR**, the above instructions would move d'30' into **Tailor**, and *not* **Tinker**.

Exercise 2.37 Given the declarations above, what *five* lines are needed to move the number from **Soldier** to **Spy**.

Some useful (but not vital) tricks

1. If you are growing tired of the lengthy **goto** instruction, you may be pleased to read that it can be abbreviated to **b**. The **b** instruction (it stands for **b**ranch) does exactly the same thing as **goto**.

Example 2.19

b	**Start**	**; goes to Start**

2. Another useful trick enables you to go to a specific part of the program, and then skip any number of instructions. This is done by adding **+1**, for example after the label, in a **goto** instruction.

Example 2.20

	goto	**Start+1**	**; goes to Start and skips the next instruction**
Start	**call**	**Init**	**; sets things up**
	bsf	**porta, 0**	**; turns on an LED**

One warning with this instruction is not to use it too frequently, and avoid large skips (e.g. +14). In such cases it is probably a good idea to simply add another label at the place you want to go to. Be wary of going back to your program and adding lines (corrections or afterthoughts, etc.), because the number of lines you need to skip may change.

Example 2.21

```
        goto   Start+1    ; goes to Start and skips the next instruction
Start   call   Init       ; sets things up
        bsf    portb, 0   ; turns on buzzer
        bsf    porta, 0   ; turns on an LED
```

If we still want the program to skip to the line where the LED is turned on, we will need to remember to change the +1 to +2. Such changes are easy to forget if your program is riddled with long skipping gotos.

This final trick was suggested by Richard George, and is a more efficient way of creating a delay or just 'killing time'. Rather than involve the TMR0, it relies on the length of an instruction cycle, so you can use the TMR0 for other tasks. I've left it until now because it requires a bit more thought than the TMR0 version, but it does take up fewer lines in the program. First you must work out how many instruction cycles your time delay requires. For example if you are using a 4 MHz crystal, and wish to wait 1 second, you calculate that an instruction is executed at (4 MHz/4 =) 1 MHz, and so you will need to 'kill' 1 million clock cycles. Now divide this number by 3, don't worry if it isn't a whole number, just round it to a whole number – the inaccuracy will be a matter of clock cycles, i.e. not even worth noting. In our example we have 333 333. We convert this number to hexadecimal (using a calculator of course!) and get 51615h (the 'h' at the end reminds me that it is a hexadecimal number). Now, write the number as an even number of hexadecimal digits (i.e. if it has an odd number of digits, stick a 0 in front), we get 051615h. Now count the number of digits this number has (it should be an even number), and write down a '1' followed by that number of 0's. In our example the number has six digits, so we write a 1 and six 0's – 1000000h. Subtract the previous number from this one: 1000000h − 051615h = FAE9EBh. Finally we split this number into groups of two hexadecimal digits, starting from the right. The result is EBh, E9h, and FAh.

At the start of the delay in the program we put these numbers into file registers:

```
        movlw   h'EB'    ; sets up delay registers
        movwf   Delay1   ;
        movlw   h'E9'    ;
        movwf   Delay2   ;
        movlw   h'FA'    ;
        movwf   Delay3   ;
```

The delay itself consists of about three lines per delay register (i.e. in our case eight lines).

```
Loop    incfsz    Delay1, f    ; this block creates a fixed delay
        goto      Loop         ;
        incf      Delay1, f    ;
        goto      Loop         ;
        incfsz    Delay2, f    ;
        goto      Loop         ;
        incf      Delay1, f    ;
        incfsz    Delay3, f    ;
        goto      Loop
```

When it finally skips out of the last Loop, 1 second will have passed. Obviously if you are concerned about the six lines which set up the delay register (first, you are a pedant!), and secondly just subtract six from your original number of instruction cycles to be wasted.

Now, I apologise for dragging you through a great deal of seemingly random arithmetic – it will now be explained. The **incfsz** instruction takes one cycle, and the **goto** takes two – so Delay1 register is incremented every *three* instruction cycles. This is why we divided the original number of instruction cycles by three. We convert to hexadecimal to ensure that when we split up the big number into groups of two digits, each group will hold a number less than 256 (i.e. a number which the PIC microcontroller can handle). You may be wondering what's the point of the lines with **incf**? These are there to make the clock cycles add up. When we are stuck in the top loop, **Delay1** is incremented *once* every *three* clock cycles. However, on the occasions where it breaks out of this loop it skips (two cycles), increments **Delay1** (one cycle), increments **Delay2** (one cycle), and loops back (two cycles). Thus, with this extra increment of **Delay1**, we create *two* increments over *six* clock cycles, and still maintain the same rate. If you work through the case where it jumps out of both loops, you will find **Delay1** is incremented *three* times over *nine* clock cycles. The final question is: 'Why do we invert the number and count up, rather than just leave it alone and count down?' The reason is saving space in the program. If we were decrementing, instead of incrementing, the shortcut instruction is **decfsz**. If we take the hexadecimal number 104h as an example, if we were decrementing it we would have split it up into 1h and 04h, and we replace the **incfsz** in the delay routine above with **decfsz**, we would get 104h, 103h, 102h, 101h, 100h, after which the **decfsz** would make the program skip, making the next number 000h, at which point the second **decfsz** would also cause a skip – completely missing out 001–0FF. The point is we don't want the program to skip on 00h, but on FFh (i.e. 104h, 103h, 102h, 101h, 100h, 1FFh -> 0FFh, 0FEh, . . .). We effectively do this by counting up instead, but in order to count the same number of clock cycles as before we need to subtract it from the 10000 . . . number. Even if you don't fully understand how this works, you can simply use the handle turning described above to get the numbers you need, and take advantage of the saved space, and the liberation of the TMR0.

Finally, this method can also be used with timing subroutines where you return if the specified time hasn't passed. If you are using more than one prescaler (after the marker) then you can use a similar method. For example if you are timing 1 minute with a 2.4576 MHz crystal, you would normally split this up into a marker of 30, and postscalers of 80 and 60, or something along those lines. With this method you still use a marker of 30 (or another appropriate number), and take the remainder (80 × 60 = 4800). This is the number to convert to hex, invert, etc. (but don't divide it by three), which in this case gives 40h and EDh. Your timing subroutine could end up looking like this:

```
Timer   movfw   Mark30      ; has one minute passed?
        subwf   TMR0        ;
        btfss   STATUS, Z   ;
        retlw   0           ; no, so returns

        incfsz  Delay1      ; (initially set to 40h)
        retlw   0           ; no, so returns

        incfsz  Delay2      ; (initially set to EDh)
        retlw   0           ; no, so returns
```

You will have had to move the correct numbers (40h and EDh) into Delay1 and Delay2, respectively, somewhere else in the program (e.g. in the Init subroutine).

Final PIC5x program – 'Bike buddy'

Our final program in this chapter on the PIC54 will tie together many of the ideas covered. It will be a mileometer (odometer) and speedometer for bicycles. The device should consist of three seven-segment displays (up to 999 kilometres recorded, and an accuracy of 0.1 kph), a toggle switch to change mode (mileage/ speed), and an input from a reed switch activated by a magnet on the wheel. This is how speed and mileage are detected. With strobing this makes a total of seven outputs for the seven-segment code (RB1 to RB7), three outputs to select the correct seven segment-display (RA0 to RA2), one input for the toggle switch (RB0), and the input for the reed switch (RA3). This makes a total of 12 I/O, which conveniently just fits on the PIC54. This leaves us with one problem. When displaying the speed, one of the decimal points of the seven-segment display will need to be on, but as we've just worked out there are no spare outputs. However, the decimal point will only need to be on when the toggle switch is selecting the speedometer mode. We can therefore directly link the toggle switch to the decimal point as shown in the circuit diagram in Figure 2.30. The flowcharts are shown in Figures 2.31a–c.

I have made this flowchart slightly more detailed than usual because this time you are expected to write the program yourself. If you break things up into the boxes described in the flowchart you should be able to manage everything with little difficulty. If you get stuck, the program I wrote is Program L in Chapter 7, but remember that the way I did something in my program may not necessarily fit in with your program as the two are likely to have some differences. You may therefore need to adapt certain sections from my program if you wish to use them.

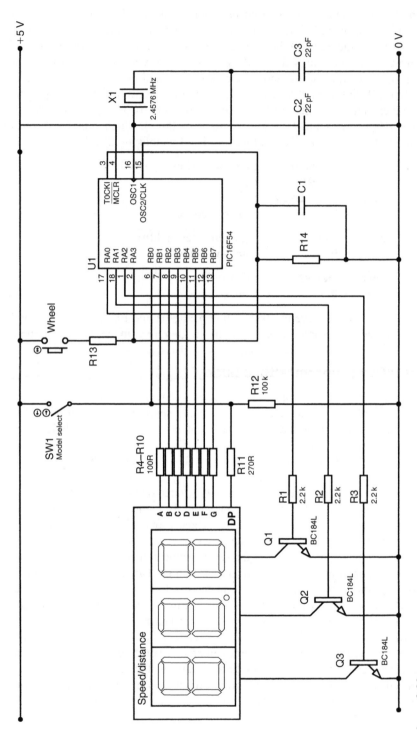

Figure 2.30

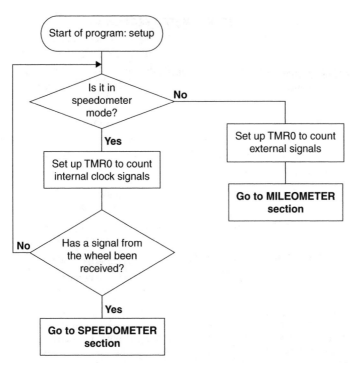

Figure 2.31a

SPEEDOMETER SECTION

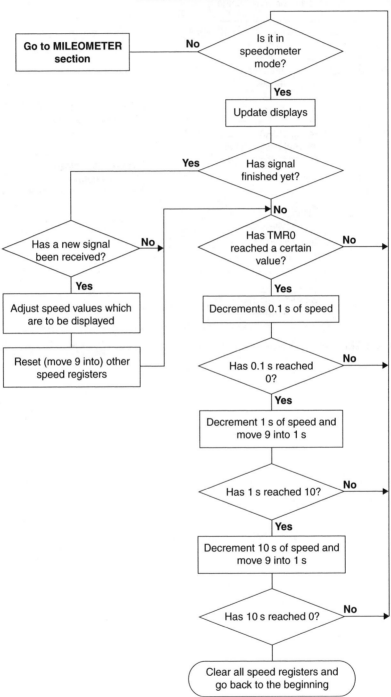

Figure 2.31b

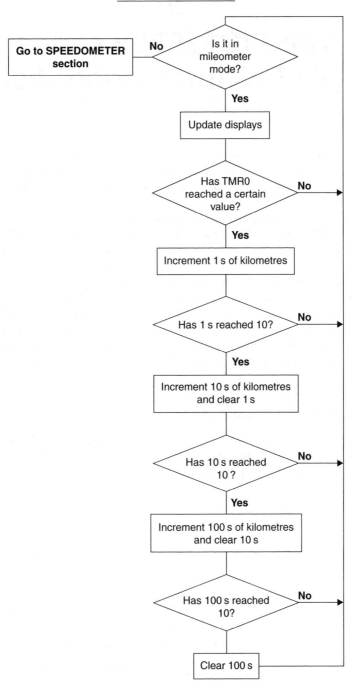

Figure 2.31c

3
The PIC12F50x series (8-pin PIC microcontrollers)

There is a range of PIC microcontrollers which manages to squeeze a large number of features into a tiny 8-pin package. The 8-pin device most like the PIC16F54 we discussed in the previous chapter is the PIC12F508 (the **12** in the name tells us that this is an 8-pin device). Surprisingly, this little PIC microcontroller offers up to 6 I/O pins (the other two are power supply pins). It needs no external oscillator (e.g. crystal or RC), as it has an in-built 4 MHz oscillator, and even offers a feature which allows external signals to wake it up from the sleep state. For any application where a small size is advantageous, and 6 I/O pins is sufficient, these kinds of PIC microcontroller are invaluable.

The PIC12F50x series consist of two models (the PIC12F**508** and PIC12F**509**) shown in Figure 3.1, with a third model (the PIC12F**510**) under development at the time of publication. The 'F509 has more memory (more program memory, and more GPFs) than the 'F508. The 'F510 will be similar to the 'F509 but with the added feature of built-in analogue-to-digital conversion (this is discussed further in Chapter 4).

Differences from the PIC16F54

There are a few differences in the way these PIC microcontrollers work, most of which are illustrated in the file registers. Figure 3.2 shows the file register arrangement for the PIC12F50x series.

The STATUS register

The first difference is found in the STATUS register. This PIC series offers the option of waking up from sleep if one of three I/O pins changes state (GP0, GP1

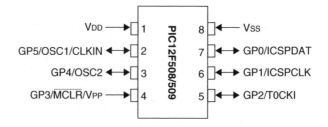

Figure 3.1

or GP3). The previously unused bit 7 of the STATUS register can now be used to see whether the PIC microcontroller was woken up from sleep due to one of these pins changing state (bit 7 is *set*), or whether it was some other reason (bit 7 is *cleared*).

The OSCCAL register

The second difference you will notice is that there is a new file register at address **05**, the **OSCCAL** file register. This is used for **os**cillator **cal**ibration, and is really only used at the start of your program (address **0x000**). To make the internal 4 MHz internal oscillator more accurate, a special number should be moved into the **OSCCAL** register. As with the PIC16F5x series, the PIC processor first executes the instruction at the last address of the program memory (1FFh for 'F508, and 3FFh for 'F509). However, when the PIC microcontrollers are made in the factory, a special instruction is programmed into them at the last address. This instruction moves a particular number (the calibration value) into the working register, i.e. it takes the form:

movlw   xx   ; moves calibration value into w. reg

After executing this instruction, the program loops back and starts at address **0x000** – remember, all this happens automatically. (By the way, if you are erasing a new PIC microcontroller, you should first read the program memory and make a note of the factory-programmed value, so that you can insert this line

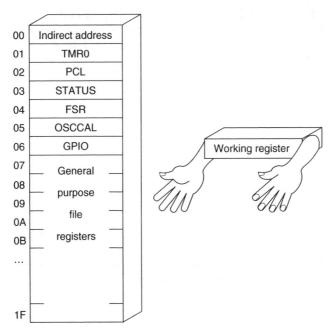

00	Indirect address
01	TMR0
02	PCL
03	STATUS
04	FSR
05	OSCCAL
06	GPIO
07	General
08	purpose
09	
0A	file
0B	registers
...	
1F	

Working register

Figure 3.2 *Map of file registers for PIC12F508.*

yourself.) If you wish to make the internal oscillator more accurate, the instruction at address **0x000** should be:

    movwf   OSCCAL   **; uses the pre-programmed value**
                                    ; to calibrate the internal oscillator

If you are not interested in oscillator accuracy, you can omit this instruction and simply place **goto start** at program at address **0x000** using the **org** command. The program template used previously should be modified as follows:

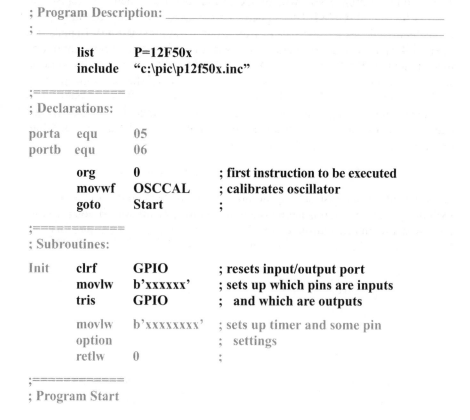

```
; Program Description: _____
; _____

        list      P=12F50x
        include   "c:\pic\p12f50x.inc"

;======================================
; Declarations:

porta   equ       05
portb   equ       06

        org       0            ; first instruction to be executed
        movwf     OSCCAL       ; calibrates oscillator
        goto      Start        ;

;======================================
; Subroutines:

Init    clrf      GPIO         ; resets input/output port
        movlw     b'xxxxxx'    ; sets up which pins are inputs
        tris      GPIO         ;  and which are outputs

        movlw     b'xxxxxxxx'  ; sets up timer and some pin
        option                 ; settings
        retlw     0            ;

;======================================
; Program Start
```

Inputs and outputs

The PIC12F50x series has only one I/O port called the **GPIO** (the general purpose input/output file register). It works in exactly the same way as Port A and Port B on the PIC54 – certain pins on the PIC microcontroller correspond to bits in this file register. One important thing to note is that *GP3 is in fact only an INPUT*, and cannot be configured as an output.

The OPTION register

As previously mentioned, the PIC microcontroller can be configured to wake up from sleep when one of GP0, GP1 or GP3 changes state. This is controlled by bit 7 of the **OPTION** register – the feature is *enabled* when bit 7 is *clear*, and *disabled* when bit 7 is *set*.

Bit 6 of **OPTION** has also been given a purpose (you may remember that these two bits were unused in the PIC54 and 57). When set, the PIC microcontroller will make pins GP0, GP1 and GP3 *float* high when not connected to anything. These are known as *weak pull-ups*. These are useful when the pins are being used as inputs which are pulled low when something happens (e.g. you've attached a push button between the pin and 0V, pulling the input low when the button is pressed). If you enable the *pull-ups* on the PIC microcontroller, you don't need an external pull-up resistor. If you don't want to use this feature then make sure you set this bit.

Note that both of these features require bits to be *set* in order to disable the feature – don't forget to do this! The rest of the **OPTION** register is as in the PIC54.

The TRIS register

Nothing much is new in this file register. Just remember that there are now 6 bits in the I/O file register, and the number you use to select inputs and outputs should reflect this. Also remember that GP3 cannot be configured as an output. Finally, note that GP2 is also the T0CKI pin. This means that if the TMR0 is configured (in **OPTION**) to count signals from the T0CKI pin, GP2 is automatically set to be an input, overriding the value of the bit in the **TRIS** register.

The general purpose file registers

The PIC12F508 is identical to the PIC54 in terms of GPFs. The PIC12F509 has an extra set at addresses 30-3Fh in the data memory. These are accessed in the same way as described for the PIC57 – by setting bit 5 of the **FSR** (see page 82).

The MCLR

The PIC12F50x series still has an $\overline{\text{MCLR}}$ pin, but if you don't need a reset pin, it can be used as an input pin (GP3). You can enable or disable the $\overline{\text{MCLR}}$ when programming the PIC microcontroller (it is one of the configuration bits). In MPLab, select 'Internal' to disable the $\overline{\text{MCLR}}$, or use the __**config** command.

Configuration bits

There are some new configuration options relating to the ability to disable the MCLR feature, and the use of the internal oscillator. Use _**MCLRE_OFF** or _**MCLRE_ON** to disable/enable the MCLR feature. The four allowed oscillator options are _**LP_OSC_**, _**XT_OSC**, _**IntRC_OSC** and _**ExtRC_OSC**, where the latter two options refer to the internal RC oscillator and external RC oscillator, respectively. An example configuration command would be:

__**config**   _**MCLRE_OFF** & _**IntRC_OSC** & _**CP_OFF** & _**WDT_OFF**

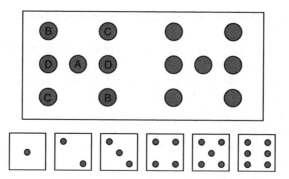

Figure 3.3

Example project: 'PIC dice'

Our example project to demonstrate the PIC12F508 will be a pair of dice, with fourteen LEDs and one button. The LEDs will be arranged as shown in Figure 3.3. When the button is pressed, the LEDs will flash randomly, and when it is released, the LEDs gradually slow down until they finally display a pair of numbers (in the traditional dice format). It will display this number for 5 seconds, then go to sleep.

The PIC12F508 supports up to five outputs, so controlling fourteen LEDs is going to be a real challenge! Looking at Figure 3.3, we notice that we don't need individual control over each LED die in order to display a number (1–6). Instead, we can split these into four groups of LEDs which I've labelled A, B, C and D. This cuts the requirement down to 8 outputs (4 per die). Finally, we can use one output to select which die is on – if the output is 0, the left die is on, and if the output is 1, the right die is on. This means we can get away with 5 outputs (1 controller, and 4 for the LEDs). The button will be connected to GP3, which will be set to wake the PIC microcontroller up from sleep. The program flowchart is shown in Figure 3.4, and the circuit diagram in Figure 3.5. As you can see from Figure 3.1, if you wish to use in-circuit serial programming, the ICSPDAT line should be connected to GP0, and the ICSPCLK line to GP1, at the programming stage. However, these pins should be disconnected from the ICSP lines during circuit operation.

In **Init** we should set up the inputs and outputs (all outputs, except GP3 which is the button). We then need to turn off all the LEDs. Looking at Figure 3.3, we see that one die's LEDs are on when their corresponding GPIO bits are 1, and the other die's LEDs are on when their corresponding GPIO bits are 0 (i.e. one has common cathode, and one common anode). Therefore to turn the LEDs off, we move **b'100000'** into GPIO. Setting Bit 5 selects the common anode group of LEDs, and so the other GPIO bits should be cleared to turn off the LEDs. Finally, set up the OPTION register with TMR0 prescaled by the maximum amount, weak pull-ups disabled, the wake-up featured enabled on pins GP0, 1 and 3.

There are three main loops in the main section of the program. In the first, we are waiting for the button to be released, the displays are randomly flashing, and

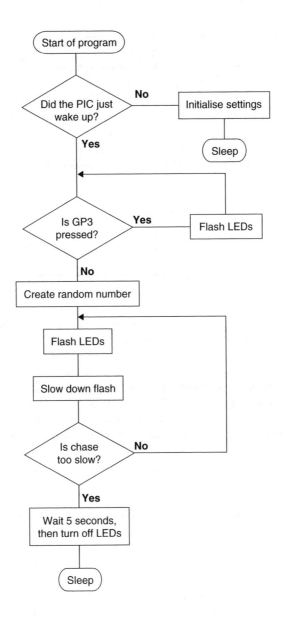

Figure 3.4

a random number is being selected. In the second, when the button is released, the displays slow down until a critical point is reached. Finally, the random number is displayed, and we wait for 5 seconds before going back to sleep.

Random digression

There are two approaches to generating random numbers: we can use some user input (e.g. the length of time a button is pressed) or another external component,

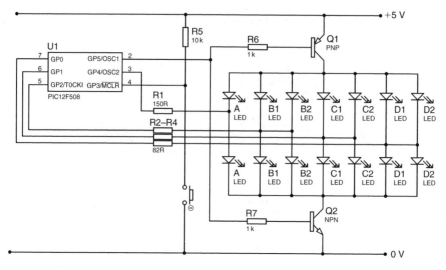

Figure 3.5

or alternatively we can use an algorithm to generate a pseudo-random number. For example, if we increment a register continually (and very quickly) during a loop in which we wait for a button to be released, the register will be overflowing constantly and will end up at a random value. If we don't have the luxury of an external input, there are methods ranging in complexity for generating random numbers. A simple algorithm is the Linear Congruential Method developed by Lehmer in 1948, and has the following form:

$$I_{n+1} = \text{mod}_m(aI_n + c)$$

This generates the next number in the sequence by multiplying the previous number by a, adding c, and taking the result modulo m. $\text{mod}_m(x)$ is equal to the remainder left when you divide x by m. Conveniently, the result of every operation performed in a PIC program is effectively given modulo 256. For example, we add 20 to 250. The 'real' answer is 270, however, the result given in a PIC program is 14. 14 *is* '270 modulo 256' or $\text{mod}_{256}(270)$. There are a number of restrictions on the choice of a and c in the above equation that maximise the randomness of the sequence. For example we could pick $a = 3$ and $c = 63$. You also have to pick a 'seed' – the first number in the sequence (I_0). You can set up this model on a spreadsheet and examine its *quasirandom* properties. First, you should notice that the randomness of the sequence does not appear to be sensitive to the seed. You should also observe that the sequence repeats itself every 256 numbers – this is an unfortunate consequence of the algorithm, but picking a larger modulus will increase the period accordingly.

  In this example project, we will use the first method (increment quickly while a button is pressed) to pick the final random number for the dice. However, for the random flashing that occurs prior to the answer being displayed, we will use the algorithm given above. The program begins:

Start	**call**	**Init**	; **initialisation procedure**
Pressed	**btfsc**	**GPIO, 3**	; **tests button**
	goto	**Released**	; **branches when released**
	call	**RandomScroll**	; **quickly increments numbers**
	call	**Timing**	; **keeps flashing going**
	call	**Display**	; **keeps displays changing**
	goto	**Pressed**	;

In this loop we wait for the button to be released. In the **RandomScroll** subroutine, the dice result (a number between 0 and 35) is incremented. This number is stored over two file registers called **Ran1** and **Ran2** which each hold a number between 0 and 5.

Exercise 3.1 What *13* lines make up the **RandomScroll** subroutine? Each time the subroutine is called, **Ran1** should be incremented – when it reaches 6 it should be reset to 0, and **Ran2** incremented. When **Ran2** reaches 6, it should be reset to 0.

The **Timing** subroutine creates the delay between the displays changing. When the button is released, this delay increases so that the dice slow down. The basic unit of time will be 1/50th of a second (hence for a 4 MHz oscillator and TMR0 prescaled by 256, we use a marker of 78). The postscaler will be set to 4 while the button is pressed (corresponding to the displays changing at a rate of about 12 times per second). When the button is released, we will set a bit called **slow**, which will tell the **Timing** subroutine to increment the postscaler up to a maximum of 31 (i.e. over the course of about 10 seconds it will slow down to a rate of about 1 per second). You can play with these values to create the type of behaviour you desire. This subroutine starts as follows:

Timing	**movfw**	**Mark78**	; **base unit = 1/50th second**
	subwf	**TMR0, w**	;
	btfss	**STATUS, Z**	;
	retlw	**0**	;
	movlw	**d'78'**	; **resets marker**
	addwf	**Mark78, f**	;
	decfsz	**PostX, f**	; **variable postscaler**
	retlw	**0**	;

PostX is the variable postscaler that is reset with a value given by **PostVal**. Thus, to slow down the flashing, we increment **PostVal**. At the point following the above code, the variable length delay has elapsed and we need to change the display values. We have a file register called **Random** containing a random number between 0 and 255 which is generated using the algorithm given above: $Random_{n+1} = mod_{256}(3\ Random_{n+63})$. This is generated by calling the subroutine **RandomGen**.

Exercise 3.2 **Challenge**: What *five* lines make up the **RandomGen** subroutine which generates a new value for the file register **Random** based on its old value.

This random number then needs to be changed into a number between 0 and 7 (as well as displaying numbers 1–6, 'all-on' and 'all-off' will be options during the random flashing). This is best done as follows:

```
swapf    Random, w    ;
andlw    b'00000111'  ; converts to 0-7 and moves
movwf    Die1num      ;  into Die1num
```

The file registers **Die1num** and **Die2num** will be used to hold the number to be displayed on the corresponding set of LEDs. Note that we do *not* simply take the 3 least significant bits of Random, as this leads to a periodicity of 8 in the random flashing, which will be very noticeable. By taking bits 4 to 6 of Random we get a period of 128, which will be much harder to spot. We use a similar set of four lines to move a random number into **Die2num**.

We then test the bit called **slow**, and call a subroutine named **Slowdown** if it is set (remember to clear it in **Init**).

Exercise 3.3 Write the *four* lines which make up the **Slowdown** subroutine which increment **PostVal** until it gets to 32, upon which it is reset to 0.

Finally, the variable postscaler **PostX** is reset with the value in **PostVal**, and we return from the subroutine.

In the display subroutine, we handle the strobing of the two sets of LEDs. Like in the Stopwatch project in the previous chapter, we use TMR0 to control strobing (in particular, bit 4 of TMR0). We'll need two look-up tables to take the number to be displayed (a number between 0 and 7 stored in **Die1num** and **Die2num**) and return the appropriate code for GPIO. '0' will correspond to all LEDs off, '1–6' correspond to the images shown in Figure 3.3, and '7' corresponds to all LEDs on.

Exercise 3.4 Write the *ten* lines which make up the **Display** subroutine. Also write the two look-up tables for the two dice (*nine* lines each). HINT: The pin arrangement for GPIO 5:0 is: Control, A, -, B, C, D, as given in Figure 3.3.

When the button is released, we jump to the **Released** section. The loop is much the same, with the exception that the **slow** bit is set, and we test for PostVal to reach 0 before skipping out of the loop:

```
Released  bsf    slow          ; tells Timing to slow down
          call   Timing        ; handles variable delays
          call   Display       ; updates displays
          movf   PostVal, f    ; has PostVal been cleared?
          btfss  STATUS, Z     ;
          goto   Released+1    ;
```

At this point, the numbers from **Ran1** and **Ran2** are incremented and moved into **Die1num** and **Die2num**, respectively. In the final loop we display the result for 5 seconds (which we create using **Mark78** and a postscaler of 250).

Exercise 3.5 Which *14* lines put the appropriate number in PostX, and then waits 5 seconds while keeping the displays going? Finally, all the LEDs should be turned off, and the PIC microcontroller should go to sleep.

When GP3 changes again (i.e. the button is pressed), the PIC microcontroller will wake up, so the **sleep** command needs to be followed with the line **goto Start**.

This completes the dice project, which gives an example of what can be achieved on the tiny 8-pin PIC microcontrollers. The full program is shown in Program M, however, note that the display codes used are dependent on how you wire up the LEDs in your circuit board, and these may not necessarily match my values. A nice extension of this project would be to change the time at which the two dice finish 'rolling', such that one finishes before the other, to create a greater air of suspense. You may also need to add some element of de-bouncing, depending on the type of button you use.

4
Intermediate operations using the PIC12F675

Studying devices such as the 'baseline' PIC5x series (by which I mean PIC16F5x and PIC12F50x chips) allows us to learn about the basics behind PIC programming. The simplicity and low cost of these entry-level devices are definite advantages; however, this also means they lack some useful features. These features include *analogue to digital conversion* (measuring an analogue voltage), *interrupts* (which save having to test inputs manually), and an *EEPROM* (a bank of data which stays intact even when you remove power). These features are all found on a rather handy little 8-pin device called the PIC12F675. It is worth noting that this is a more 'typical' kind of PIC microcontroller (rather than the simple PIC5x series) and so if you come across a new PIC microcontroller it is more likely to behave like this one. If you decide that 6 I/O pins are too few, there is a 14-pin version called the PIC16F676 which is essentially identical to the PIC12F675 but has 12 I/O pins.

Looking at the pin layout of the PIC12F675 in Figure 4.1, you should notice similarities and differences between it and the PIC12F508 of the previous chapter. You will also see that some of the pins are labelled AN0, AN1, AN2 and AN3: these can be made analogue inputs. VREF (pin 6) can be made the voltage reference for the other analogue inputs (i.e. the PIC microcontroller compares the voltage at the other pins with the voltage on the VREF pin). INT (pin 5) can be set to interrupt normal program flow when it goes high or low. The pins labelled CIN+, CIN− and COUT are part of a *comparator* module. A comparator compares the voltage on two inputs, and tells you which one is greater. Finally, this PIC microcontroller has not one timer but two! The second is called TMR1 (in addition to the TMR0 we have been using). The pins labelled T1CKI and $\overline{\text{T1G}}$ are associated with this second timer.

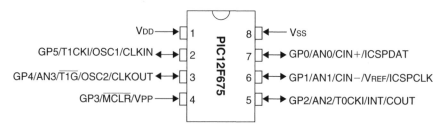

Figure 4.1

Due to the compact nature of the PIC12F675, many of these different pin functions are squeezed onto the same pins and we often have to make a choice of which particular function we wish to use. On larger models these features are spread over more pins and we have more choice over which ones can be used at the same time. Each of the pins described above, and their associated features, will be covered in detail in this chapter.

The inner differences

Having looked at the external differences, we now need to examine the inside of this PIC microcontroller. Figure 4.2 shows the arrangement of file registers on

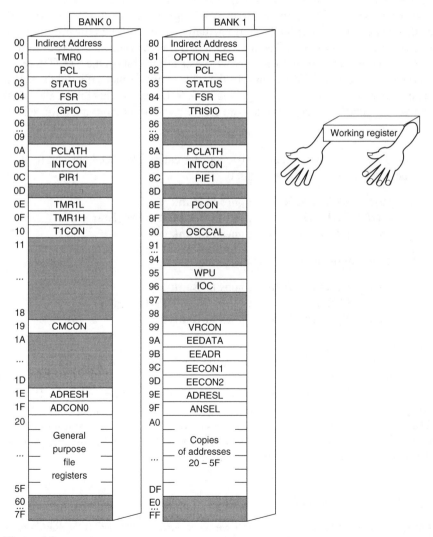

Figure 4.2

the PIC12F675. The first thing to observe is that all the extra features bring with them a load of extra special function registers (SFRs). Do not be overwhelmed by the large quantity of these SFRs – we will go over each one in due course. The greyed file registers are unused areas of the data memory. If you try reading the values in these locations, you will get a 0.

The second thing you might notice is that there are two *banks*. Whereas the PIC16F54 had only one bank ('filing cabinet'), the PIC12F675 has two sets of file registers. You should also take note that some file registers are the same in Bank 1 as in Bank 0. Think of a bank as a 'frame of mind' of the PIC microcontroller, where file registers may (or may not) be different depending on the 'frame of mind'. File register 03 will always be the STATUS register, regardless of the 'frame of mind' the PIC microcontroller is in. However, in Bank 0, file register 05 will be GPIO, and in Bank 1 file register 05 actually corresponds to a file register called TRISIO. Even if I actually write 'GPIO' in the program, the PIC microcontroller will still act on TRISIO, if it is the Bank 1 'frame of mind'.

To switch from one bank to another we use one of the bits in the STATUS register (now you see why STATUS must be the same in both banks – if it didn't exist in Bank 1 there would be no way of getting back to Bank 0!). This bit is called RP0 and is bit number 5. To go to Bank 1, we set the bit. To return to Bank 0 we clear it.

Example 4.1 We want to clear the file register called TRISIO, however, the PIC microcontroller is currently in Bank 0.

        bsf   STATUS, RP0   ; goes to Bank 1
        clrf   TRISIO   ; clears the TRISIO register
        bcf   STATUS, RP0   ; goes to Bank 0

Note that the following performs the same task:

        bsf   STATUS, RP0   ; goes to Bank 1
        clrf   GPIO   ; clears the TRISIO register
        bcf   STATUS, RP0   ; goes to Bank 0

Naturally writing 'TRISIO' makes far more sense – but the point is to highlight the fact that if you try to do something to GPIO when in Bank 1, you will actually do it to TRISIO.

In many cases, a Bank 1 file register is in some way related to its Bank 0 counterpart (e.g. the OPTION_REG register is largely a setup register for TMR0). Because the Bank 1 file registers tend to be involved in setting up, you may only need to go into Bank 1 during the **Init** subroutine. Finally, please note that the PIC microcontroller starts up in Bank 0.

The OPTION and WPU registers

From the top, the first new file register we come across is the OPTION_REG register. It isn't strictly a new file register, because there was an OPTION register on the PIC5x chips, however we did not have direct access to it. Remember how to move a number into the OPTION register (e.g. in order to set up TMR0) with the PIC5x? Below is a reminder:

> **movlw**   **b'xxxxxxxx'**   **; moves the number into w. reg**
> **option**   **; moves w. reg into OPTION**

With most other PIC microcontrollers (including the PIC12F675) there is no need for this **option** instruction as we can simply move the number into the OPTION register as we would with any other:

> **movlw**   **b'xxxxxxxx'**   **; moves the number into w. reg**
> **movwf**   **OPTION_REG**   **; moves the w. reg into OPTION**

First you should note that we use the term **OPTION_REG** to describe the OPTION register in the program – this distinguishes it from the (now defunct) **option** instruction. Secondly, I should remind you that if you don't switch into Bank 1 before you perform the above two lines, you will actually move the number into TMR0.

Bit 7 of the OPTION register now controls internal *pull-ups*, which are available on all the I/O pins (expect GP3). As before, when the bit is *set* all the pull-ups are totally *disabled*, and when this bit is *clear*, the pull-ups are *enabled* (in general). If pull-ups are enabled in general, then they can be individually enabled or disabled using the WPU register (**W**eak **P**ull-**u**p register). Each bit in the WPU controls the correspond bit in GPIO (e.g. setting bit 0 of WPU *enables* the pull-up on bit 0 of GPIO, and clearing bit 4 of WPU *disables* the pull-up on bit 4 of GPIO).

Bit 6 of the OPTION register is associated with interrupts and will be discussed later. The remaining bits of the OPTION register are the same as before.

The TRISIO register

The same new method applies to writing to the TRIS file register. Rather than using the **tris** instruction (which doesn't exist on this PIC microcontroller), we can move the number directly into TRISIO (again, this has to take place when in Bank 1):

> **movlw**   **b'xxxxxx'**   **; moves a number into w. reg**
> **movwf**   **TRISIO**   **; sets up inputs and outputs on GPIO**

Calibrating the internal oscillator

Finally, if we wish to use the 4 MHz internal oscillator we need to calibrate it (as we did with the PIC12F508). There are a few important differences to note:

1. The reset vector of this device is **0x000** (i.e. it starts at the beginning of the program memory).

2. The program memory is no longer split into pages. We have the freedom to **goto** or **call** to anywhere without worrying about page bits.
3. OSCCAL is now in Bank 1.
4. The following calibration instruction has been placed at address **0x3FF** (the last address of program memory):

> **retlw   XX**   ; **returns with calibration value in w. reg**

Therefore, the code to set up the internal oscillator should now be placed in the **Init** subroutine and consists of:

> **bsf   STATUS, RP0**   ; **goes into Bank 1**
> **call   3FFh**   ; **calls calibration address**
> **movwf OSCCAL**   ; **moves w. reg into OSCCAL**
> **bcf   STATUS, RP0**   ; **goes back to Bank 0**

After executing the line **call   3FFh**, the program returns with the factory-programmed calibration value in the working register, which is then moved into OSCCAL.

PCLATH: Higher bits of the program counter

While PCL holds the lower eight bits of the program counter (bits 0 to 7), the higher bits are not directly accessible. With the PIC5x series we had some handle on the higher bits using *page* bits in the STATUS register. On the PIC12F675 these page bits are largely unnecessary, but are effectively stored in PCLATH. You don't need to worry about the upper bits of the program counter during **goto**s and **call**s, however you have to be careful when doing variable jumps (i.e. adding numbers to the program counter). When you do this, as well as performing the operation on the PCL, the PIC processor will load the state of PCLATH into the upper bits of the program counter (PCLATH feeds directly into the upper byte of the PC). For example, if I have a lookup table which starts at address **0x240**, I need to move **2** into PCLATH before adding anything to the PCL. In the example below, the lookup table starts at address **0x045** so we need to clear PCLATH first.

Example 4.2

0045	clrf	PCLATH	; makes sure PCLATH is 0
0046	movfw	Marx	; reads in value from file
0047	addwf	PCL, f	; adds to PCL for variable jump
0048	goto	Groucho	; branches accordingly
0049	goto	Harpo	;
0050	goto	Chico	;

Remaining differences

The remaining new SFRs can be divided into a number of categories, which will be dealt with in turn:

INTCON, PIR1, PIE1, IOC:	**Interrupts**
EEDATA, EEADR, EECON1, EECON2:	**EEPROM**
CMCON, VRCON:	**Comparator**
ADRESH, ADRESL, ADCON0, ANSEL:	**Analogue to Digital Conversion**
TMR1L, TMR1H, T1CON:	**Timer 1** (a second timer)

The PIC12F675 also boasts a stack which is *8 levels* deep (compared with 2 levels deep on the PIC5x series). This means you can call a subroutine within a subroutine within a subroutine within a subroutine . . . etc., etc.! Having the third level is particularly useful; the others may not be used that often.

There are two more instructions found on the PIC12F675 and most other PIC microcontrollers (but not on the PIC5x series):

    addlw   number ;

(Not for PIC5x series) – **add**s a literal (**number**) to the number in the working register.

    sublw   number ;

(Not for PIC5x series) – **sub**tracts the number in the working register from a literal (**number**), leaving the result in the working register.

Finally, note that the watchdog timer (WDT) timeout behaves slightly differently when this PIC microcontroller is in sleep mode. Rather than causing a full reset, as on the PIC5x series, a WDT timeout during sleep causes this PIC microcontroller to wake up, and continue executing the program from the line after the **sleep** command. When not in sleep, a WDT timeout causes a full reset, as usual.

Interrupts

An *interrupt* tells the PIC microcontroller to drop whatever it's doing and go to a predefined place (the *interrupt service routine* or *ISR*) when a certain event occurs. Think of it as a fire alarm which goes off when something is detected, and makes the PIC microcontroller go to a particular meeting point. This event could be receiving a signal on the INT (GP2) pin, or perhaps the state of one of the other I/O pins changing. An interrupt can be set to occur when one of the timers (TMR0 or TMR1) overflows, and there are interrupts associated with the EEPROM, analogue to digital converter and the comparator. Each of these interrupts can be enabled or disabled individually, and many can be active at the same time. As they all interrupt the program and make the program jump to the same place (the *ISR*), you may be wondering how we can tell which event triggered caused the interrupt. Fortunately, as well as having individual enable bits, each interrupt also has an associated flag which can be tested to see if that particular interrupt has occurred. At the start of the ISR you should test the flags of all enabled interrupts and branch off to difference sections accordingly. Note also that these interrupt flags *must be cleared by you*, so somewhere during the ISR you should clear the flag so it's ready to trigger next time. Finally, note that interrupt flags will get set *regardless of the state of the interrupt enable* – the interrupt enable only dictates whether or an interrupt flag going high will actually trigger an interrupt.

The majority of the interrupt enable bits and flags are held in the **INTCON** (**Int**errupt **Con**trol) register. A few further interrupts, known as 'peripheral interrupts' have individual enable bits in the **PIE1** (**P**eripheral **I**nterrupt **E**nable) register, and flags in **PIR1** (**P**eripheral **I**nterrupt **R**egister). Let's start with INTCON:

INTCON

Bit no.	7	6	5	4	3	2	1	0
Bit name	**GIE**	**PEIE**	**T0IE**	**INTE**	**GPIE**	**T0IF**	**INTF**	**GPIF**

Port Change flag
1: A GPIO change interrupt occurred
0: It hasn't
[**Note:** Must be cleared by you]

External INT flag
1: An INT (GP2) interrupt has occurred
0: It hasn't
[**Note:** Must be cleared by you]

TMR0 Overflow Interrupt flag
1: TMR0 has overflowed
0: TMR0 has not overflowed
[**Note:** Must be cleared by you]

Port Change Interrupt Enable
1: Enables GPIO port change interrupt
0: Disables it

External INT Interrupt Enable
1: Enables the INT (GP2) interrupt
0: Disables it

TMR0 Overflow Interrupt Enable
1: Enables TMR0 overflow interrupt
0: Disables it

Peripheral Interrupt Enable
1: Enables any enabled 'peripheral interrupts'
0: Disables all 'peripheral interrupts'

Global Interrupt Enable
1: Enables any enabled interrupts
0: Disables ALL interrupts

Bit 7 (GIE) is the global interrupt enable, which is the master switch for all interrupts. Turn it off and no interrupts are enabled (regardless of the state of their individual enable bits). Turn it on and interrupts whose individual enable bits are set will be enabled.

Bit 6 (PEIE) is a mini-master switch for a group of interrupts which are known as 'peripheral interrupts'. These interrupts have their own enable bits in the PIE1 register. Therefore, in order to use these interrupts you have to enable three bits – the individual enable bit in PIE1, this bit, and the global interrupt enable.

Set bit 5 (T0IE) to use the TMR0 overflow interrupt – this simply triggers an interrupt whenever TMR0 overflows from 255 to 0. In the interrupt service routine you can test bit 2 (T0IF) to see if a TMR0 overflow interrupt has occurred (remember that you need to clear it yourself!).

Bit 4 (INTE) controls the 'External Interrupt' which depends on the state of the pin labelled INT (GP2). The interrupt can be set to trigger on the rising edge *or* falling edge of the signal on this pin. This is done using bit 6 of the OPTION register: if bit 6 of OPTION is *clear*, the INT interrupt will occur on the *falling* edge of the INT pin. If bit 6 of OPTION is *set*, the INT interrupt will occur on the *rising* edge.

Finally, bit 3 (GPIE) of the INTCON register controls the GPIO change interrupt. This interrupt can trigger when any one of the GPIO pins *changes*. To use this interrupt you need to set this bit, and also select which GPIO pin should be able to trigger the interrupt. This is done with the **IOC** (Interrupt **O**n **C**hange) register. Each bit in the IOC corresponds to a bit in GPIO – set the bit to enable interrupts when that pin changes. For example, to enable an interrupt to occur whenever pins GP0, GP2 and GP4 change, you should write the following:

```
bsf     STATUS, RP0   ; moves into Bank 1
movlw   b'00010101'   ; enables GP0, GP2 and GP4
movwf   IOC           ;  for the GPIO change interrupt
movlw   b'10001000'   ; enables GPIO change interrupt,
movwf   INTCON        ;  and enables global interrupts
```

The interrupt service routine

When an interrupt takes places, the PIC processor will jump to the instruction at address **0x004**. What's more, it actually *calls* a subroutine which starts at address **0x004**. This is so that after dealing with the interrupt, the processor can return to where it left off before the interrupt occurred. In our previous programs, address **0x004** has been five lines into our **Init** subroutine, so we will have to make some changes to the template. At address **0x004** we want to **goto** somewhere which we will call **isr**. When the processor comes across the return instruction in **isr**, it will return to the point in the program which it was at when the interrupt occurred. Remembering that the reset vector for the PIC12F675 is **0x000**, we could write:

```
org     0
goto    Start
```

```
            org     4
            goto    isr
Init        etc.
```

The only problem with this is that you are wasting addresses **0x001** to **0x003**, however this is not serious. Alternatively, you could write the following:

```
            org     0
            goto    Start

Init        clrf    GPIO          ;
            movlw   b'xxxxxxxx'   ;
            goto    InitCont      ; skips address 0x004
            goto    isr           ; at address 0x004 goes to isr
InitCont    etc.                  ; carries on with rest of Init
```

Counting down you should see that the line **goto   isr** is still at address **0x004**, and rather than losing three lines (**0x001** to **0x003**), we really only waste one line (**goto   InitCont**).

The start of the interrupt service routine should begin by checking which particular event triggered the interrupt (if more than one input is enabled).

```
isr         btfss   INTCON, 0   ; did GPIO change interrupt occur?
            goto    GPchange    ; yes, it was a GPIO change
            btfss   INTCON, 1   ; did the INT/GP2 occur?
            goto    External    ; yes, it was the INT interrupt
            btfss   INTCON, 2   ; did the TMR0 overflow?
            goto    Timer       ; yes, it was the TMR0 interrupt
            etc.
```

Fortunately, the processor automatically clears the global interrupt enable bit (GIE) in the INTCON register when an interrupt occurs. This means that no interrupt can take place in the ISR – you can imagine the havoc that would take place should this not be the case! Thus, at the end of the ISR we would have to set the global interrupt enable just before returning, but even if we did this, an interrupt could take place immediately afterwards, before actually returning from the ISR. We can't set the global enable *after* returning because we don't know where the processor is going to return to. Fortunately there is a new instruction which solves this problem:

```
        retfie              ;
```

This **ret**urns **f**rom a subroutine and sets the global **i**nterrupt **e**nable bit *at the same time*. In certain cases you may want to return from the ISR (or indeed any subroutine) without setting the global interrupt enable. On the PIC5x series, **retlw** is the only available instruction. On the PIC12F675 we can use:

```
        return              ;
```

This simply **return**s from a subroutine.

Interrupts during sleep

If an interrupt which has been individually enabled occurs during sleep, the PIC microcontroller will wake up and do one of two things, depending on the state of the global interrupt enable (GIE). If the GIE is off, it will just wake up from sleep and carry on running through the program from the line after the **sleep** command. If the GIE is set, the processor will execute the instruction after **sleep**, and then call the ISR (address **0x004**). Therefore, if you just want to use an interrupt to wake up the PIC microcontroller, you should clear the GIE *before* the **sleep** instruction. If you want the program to respond in some other way, you should make sure GIE is set. Note that the TMR0 is off during sleep, so the TMR0 interrupt cannot be used to cause a wake-up from sleep.

Example 4.3 Make the PIC microcontroller go to sleep until triggered by a change of state of inputs GP0 or GP1 (assume these two have already been enabled in the **IOC** register). It should then carry on with the rest of the program with the TMR0 and GPIO change interrupts enabled.

```
movlw    b'00001000'      ; only enables GPIO change interrupt
movwf    INTCON           ;  and disables GIE
sleep                     ; goes to sleep
movlw    b'10101000'      ; enables TMR0, GPIO change, and
movwf    INTCON           ;  global interrupts
```

Exercise 4.1 Write the *seven* lines to send the PIC microcontroller to sleep, and be woken by the *rising* edge of the INT (GP2) pin. Upon waking, the program should *do nothing* before calling the ISR. (**Hint**: Don't forget to configure the relevant bit in the OPTION register.)

That's all there is to interrupts; just remember to make the ISR fairly short, because you can't get an interrupt while you're in it. Think clearly when writing this part of the program, particularly if you have more than one interrupt enabled.

Maintaining the STATUS quo

Remember that interrupts can occur at any point during the program. We could be moving something into the working register, and be about to move it into another file register when *WHAM!*, an interrupt occurs. When we return from the ISR, there is likely to be a new number in the working register – what happens now?

```
Example 4.4  movlw    d'15'            ; has MinutesFame reached 15?
             subwf    MinutesFame, w   ;
             btfss    STATUS, Z        ;
```

In Example 4.4, what happens if an interrupt occurs after the second line? Upon returning from the ISR, the zero flag may be in a different state. In order to ensure that interrupts don't disrupt the functioning of the program, we have to store the contents of the working register and the STATUS register at the beginning of the ISR. At the end of the ISR we copy these values back and then return. To store the original values we use:

```
movwf   W_temp          ; stores w. reg in temp register
movfw   STATUS, w       ; stores STATUS in temp
movwf   STATUS_temp     ; register
```

And to restore the two registers at the end of the ISR we use:

```
movfw   STATUS_temp     ; restores STATUS register to
movwf   STATUS          ;  original value
swapf   W_temp, f       ; restores working register to
swapf   W_temp, w       ;  original value
```

This may seem a little puzzling. Why not simply move **W_temp** directly into the working register using the **movfw** command? The reason is that the **movfw** instruction affects the zero flag, and so has the potential of altering the original value of the STATUS register. Fortunately, the **swapf** instruction does not affect the zero flag, and so is suitable in this case. Note that *swapping* twice results in no net change to the value, and so these two instructions move the value from **W_temp** into the working register with no change to STATUS.

New program template

With all these new file registers it is clear that our program template needs to be updated. A good practice is to clear any control file registers that you are not using. The only exception to this is the comparator module, which is in a low-power mode if the **CMCON** (**c**omparator **m**odule **con**trol) register is clear, but is turned completely off if you *set* bits 0–2, as shown in the template below. Even if you are going to use interrupts in the program, you should *not* set the global interrupt enable until everything else is configured. You can then use the **retfie** instruction to leave **Init** and enable global interrupts. If you don't want to enable interrupts at this point, end **Init** with the **return** instruction. If you are not using interrupts, you can remove the ISR.

```
;***********************************
; written by:                     *
; date:                           *
; version:                        *
; file saved as:                  *
; for PIC ...                     *
; clock frequency:                *
;***********************************
```

```
; Program Description: _____
; _____

                list     P=12F675
                include  "c:\pic\p12f675.inc"

;============
; Declarations:

W_temp          equ      20 h
STATUS_temp     equ      21 h

                org      0              ; first instruction to be executed
                goto     Start          ;

                org      4              ; interrupt service routine
                goto     isr            ;

;============
; Subroutines:

Init            bsf      STATUS, RP0    ; goes to Bank 1
                call     3FFh           ; calls calibration address
                movwf    OSCCAL         ; moves w. reg into OSCCAL
                movlw    b'xxxxxx'      ; sets up which pins are inputs
                movwf    TRISIO         ;   and which are outputs
                movlw    b'xxxxxx'      ; sets up which pins have
                movwf    WPU            ;   weak pull-ups enabled

                movlw    b'xxxxxxxx'    ; sets up timer and some pin
                movwf    OPTION_REG     ;   settings
                clrf     PIE1           ; turns off peripheral ints.
                clrf     IOC            ; disables GPIO change int.
                clrf     VRCON          ; turns off comparator V. ref.
                clrf     ANSEL          ; makes GP0:3 digital I/O pins

                bcf      STATUS, RP0    ; back to Bank 0
                clrf     GPIO           ; resets input/output port
                movlw    b'00000111'    ; turns off comparator
                movwf    CMCON          ;
                clrf     T1CON          ; turns off TMR1
                clrf     ADCON0         ; turns off A to D conv.
                movlw    b'0xxxxxxx'    ; sets up interrupts
                movwf    INTCON         ;

                retfie or return        ;

isr             movwf    W_temp         ; stores w. reg in temp register
                movfw    STATUS         ; stores STATUS in temporary
                movwf    STATUS_temp    ;   register
```

(Write the interrupt service routine here)

movfw	**STATUS_temp**	**; restores STATUS register to**
movwf	**STATUS**	**; original value**
swapf	**W_temp, f**	**; restores working register to**
swapf	**W_temp, w**	**; original value**
retfie or **return**		**; returns, enabling GIE**

;=============
; **Program Start**

Start	**call**	**Init**	**; initialisation routine**
Main	(Write your program here)		
	END		

Example project: 'Quiz game controller'

The project to practice interrupts will be a quiz game device. There will be three push buttons (one for each player), three LEDs (one by each button to show which player pressed first), and a buzzer to show that a button has been pressed which stays on for 1 second. There will also be a button for the quizmaster to reset the system (this can be connected to the GP3/$\overline{\text{MCLR}}$ pin). You may wonder why we are going to the trouble of using interrupts for this project, which looks as if it may be viable on the PIC16F54. However, without interrupts we would have to test each button in turn, one after the other. Let us say, for example, that the program had just finished testing the first button, and then immediately afterwards, the first button is pressed. The program then tests the second button, after which the third player responds. The third player's button is now tested, and as far as the program is concerned, he responded first. The times we are dealing with are millionths of a second, but if we want to be really exact we can use interrupts. The three players' buttons could be connected to pins which have the GPIO change interrupt enabled, so that this interrupt would trigger the moment any button is pressed. The circuit diagram for this project is shown in Figure 4.3. Because the PIC12F675 does not exactly have an abundance of I/O pins, we have to double up the button pins and LED pins. In order for this to work, we need an extra pin (GP0) to act as a master switch for the LEDs. With this output set (+5 V), pins GP1, 2 and 4 can be made outputs to control whether the LEDs are on or off, irrespective of whether a button is pressed. When this pin is clear, pins GP1, 2 and 4 can be made inputs (with weak pull-ups enabled) which read the state of the buttons. We could avoid this complication by moving to a larger device (such as the PIC16F676), but this illustrates what can be achieved with a relatively small number of pins. In summary, the two modes of operation with regards to pin settings are:

1. *Waiting for button to be pressed:* GP0 is an output, and off (so LEDs are disabled). GP1, 2 and 4 are inputs, with weak pull-ups enabled.
2. *Displaying correct LED:* GP0 is an output, and on (so LEDs are enabled). GP1, 2 and 4 are outputs.

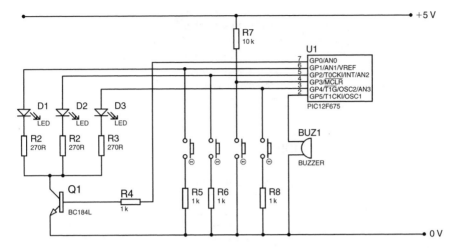

Figure 4.3

The buzzer will be connected to pin GP5. The flowchart is shown in Figure 4.4 – as you can see, the main body of the program is nothing at all, just a constant loop. All the clever stuff happens in the interrupt service routine.

Exercise 4.2 With the help of the program template above, write the **Init** subroutine for this program. We will be using the GPIO change interrupt to determine when a button is pressed, and the TMR0 interrupt to help with the timing for the buzzer. However, we don't enable the TMR0 interrupt just yet – we will do it later. Also, don't enable the global interrupt until the last line of the subroutine – **retfie**. Set up inputs and outputs to prepare for waiting for a button to be pressed, and don't forget to enable weak pull-ups on GP1, GP2 and GP4. Set up the IOC register correctly to generate interrupts-on-change for GP1, GP2 and GP4.

The main body of the program is just a loop, waiting for the GPIO change interrupt to occur. The program, from **Start**, is therefore:

Start   call   Init   ; sets everything up
Main   goto   Main   ; keeps looping

This leaves the **isr** to complete. In this project, we are using two interrupts, so we need to check the interrupt flags to determine which interrupt occurred. Note that in this particular project, it isn't essential to include the code at the start and end of the **isr** which store and recover the contents of the working and STATUS registers, as nothing is happening while we're waiting for an interrupt.

Exercise 4.3 What *two* lines are required at the start of **isr** to determine which interrupt occurred? If the GPIO interrupt occurred, continue, otherwise jump to a section called **Timer**.

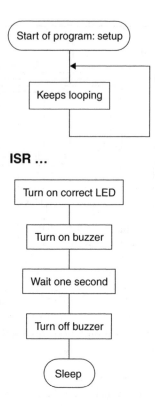

Figure 4.4

Interrupt flags need to be reset in the program, otherwise, upon returning from the ISR, the same interrupt will trigger again. The GPIO interrupt flag is reset by clearing bit 0 of INTCON. We now need to record which button was pressed, and turn on the corresponding LED. The buttons are *active low*, so the bit goes to 0 when a button is pressed. To turn this into an *active high* signal, we invert the state of GPIO, moving the result into the working register, and then mask all bits except the ones we're interested in: GP1, 2 and 4.

```
bcf     INTCON, 0   ; resets GPIO interrupt flag
comf    GPIO, w     ; inverts state of GPIO
andlw   b'010110'   ; masks all except GP1, 2 and 4
movwf   temp        ; stores result
```

The number in **temp** should now be all 0s, with a 1 at the bit corresponding to the button which was pressed. This number can then be used to turn on the correct LED. As a safety precaution to guard against problems like button bounce, etc. we can do a quick check; if the number in **temp** is 0, this was a false alarm and we should ignore it. The zero flag would have been triggered by the **andlw** instruction, and so we can test it immediately:

> **btfss**   **STATUS, Z**   **; is a button actually pressed?**
> **retfie**   **; no – false alarm, so returns**

Now we are sure a button was pressed, and **temp** holds a number corresponding to which one it was. We now need to change pins GP1, 2 and 4 to outputs (which automatically disables weak pull-ups. The contents of **temp** can then be moved back into GPIO, which will turn on the LED corresponding to the button that had been pressed. We also need to enable the LEDs by setting GP0, and turn on the buzzer by setting GP5. This can be achieved through separate **bsf** commands, but we do the same by adding the number **b'100001'** to **temp**, before moving it into GPIO.

Exercise 4.4 What *seven* lines turn GP1, 2 and 4 to outputs, and then use the number in **temp** to turn on the correct LED and the buzzer? Don't forget to move in and out of Bank 1.

We should then disable the GPIO change interrupt, and enable the TMR0 interrupt, and return from the **isr**, enabling global interrupts.

Exercise 4.5 What *three* lines complete the GPIO change part of the **isr**?

We will use the TMR0 interrupt to time the 1 second delay period for the buzzer. If we use the 4 MHz internal oscillator, instructions are executed at a frequency of 1 MHz, and TMR0 counts up at a frequency of 3.9 kHz. The frequency of the TMR0 interrupt is therefore 15.3 Hz, so if we want to time an approximate one second delay, we should use a postscaler of 16. We set up a file register with the number 16 (do this in **Init**), and decrement it each time the TMR0 interrupt occurs. After the 16th interrupt, the buzzer is turned off and the PIC microcontroller goes to sleep. Upon going to sleep, the states of the outputs stay the same, so the correct LED stays on. Before going to sleep, the program should set up INTCON so that all the interrupts are disabled. Therefore, only a reset on the $\overline{\text{MCLR}}$ pin (to which the quiz-master's button is connected) will wake up the device.

Exercise 4.6 What *six* lines make up the section called **Timer**.

This completes the program, which is given in Program M. So far we have looked at two interrupts, the GPIO change and TMR0 interrupts. Remaining interrupts on this PIC microcontroller (external interrupt on the INT pin, A/D conversion interrupt, EEPROM write interrupt and comparator interrupt) will be dealt with in subsequent parts of this chapter.

EEPROM

EEPROM (**E**lectrically **E**rasable **P**rogrammable **R**ead-**O**nly **M**emory) can be seen as a large collection of general purpose file registers whose contents remain intact even after power has been removed. We used the analogy of a filing cabinet to describe the file registers. When the PIC microcontroller is turned off, the filing cabinets are left exposed and there is little guarantee that the numbers in the file registers will be intact when you turn it on again. The EEPROM behaves like the office safe – a secure place to store data which will not be affected by removing power.

We can define this in more rigorous terms and say that the 64 file registers in the data memory are RAM (**r**andom **a**ccess **m**emory). In addition to these, there are 128 data locations which are ROM (**r**ead-**o**nly **m**emory) – the EEPROM. Reading and writing to these secure locations requires a bit more effort than with the file registers in the RAM.

The file register **EEADR** holds the **address** in the **EEPROM** which you wish to read or write to, while **EEDATA** holds the **data** that you have just read, or which you wish to write to the **EEPROM**. EECON1 holds settings for the EEPROM, and EECON2 is a special register used in the EEPROM writing process. Note that all these EEPROM file registers are found in Bank 1.

EECON1

bit no.	7, 6, 5, 4	3	2	1	0
bit name	unused	WRERR	WREN	WR	RD

Read Control Bit
1: Starts an EEPROM read
0: EEPROM read has finished

Write Control Bit
1: Starts an EEPROM write operation (stays high until write operation finishes)
0: EEPROM write has finished

EEPROM Write Enable Bit
1: Allows writing to the EEPROM
0: Forbids writing to the EEPROM

EEPROM Write Error Flag
1: An EEPROM write has prematurely terminated
0: The write operation completed without error

Reading from the EEPROM

Let's pretend for the moment that you have already written something to the EEPROM and you now wish to read it.

Example 4.5 You wish to read the number stored in address **4Eh** of the EEPROM and move it into the working register.

bsf	STATUS, RP0	; go to Bank 1
movlw	4Eh	; selects EEPROM address
movwf	EEADR	;
bsf	EECON1, 0	; starts EEPROM read operation
		; storing result in EEDATA
movfw	EEDATA	; moves read data into w. reg

After moving into Bank 1, the next two instructions tell the PIC microcontroller which address in the EEPROM you wish to read. The EEPROM Read bit is set to initiate a read from the EEPROM, putting the result in EEDATA. This file register can be read directly immediately after the read command.

Writing to the EEPROM

Writing to the EEPROM is made slightly more complicated by the fact that it is a more 'dangerous' operation. Reading from the EEPROM is quite harmless – all you are doing is moving a number into EEDATA (photocopying some documents that are in the safe). Writing to the EEPROM, on the other hand, involves actually *changing* the data in the EEPROM (altering the documents in the safe). Because of this distinction, steps are taken to minimise the risk of accidentally writing to the EEPROM. You have to provide a type of 'combination for the safe' in the program before you are allowed to write to the EEPROM.

Example 4.6 You wish to write the decimal number **69** into the EEPROM address space **78h**. First ensure the write enable bit (EECON1, bit 2) is set, then provide the 'safe combination' – a series of four instructions which must *immediately* precede the write operation. As the execution of this procedure must not be interrupted, the global interrupt enable should be cleared for the duration of the write operation.

bsf	STATUS, RP0	; goes to Bank 1
movlw	d'69'	; moves the number to be written, into
movwf	EEDATA	; EEDATA
movlw	78h	; moves the address to be written to
movwf	EEADR	; into EEADR
bsf	EECON1, 2	; enables a write operation
bcf	INTCON, 7	; disables global interrupts
movlw	55h	; now follows the 'safe combination'
movwf	EECON2	;
movlw	AAh	;
movwf	EECON2	;
bsf	EECON1, 1	; starts the write operation
etc.		

There is still a little more to the writing operation, because although we have started the write, it will take quite a few clock cycles to complete. This is in contrast to the read operation which takes place immediately. If there is something in particular we want to do when the write finishes we can wait until the write completes by testing EECON1, 1 (the write control bit) which gets cleared when the write operation finishes:

EELoop
      btfsc   EECON1, 1 ; has write operation finished?
      goto   EELoop   ; no, still high, so keeps looping

If we don't want to get tied up in some loop, but want to be able to get on with other things, we can use the *EEPROM Write Complete* interrupt, which (as you may have guessed) triggers when the write operation finishes. This interrupt is a so-called 'peripheral interrupt' and can be enabled using bit 7 of the PIE1 register. Don't forget that the peripheral interrupt enable (INTCON, bit 6) as well as the global interrupt enable need to be set in order for the interrupt to occur.

Exercise 4.7 Write the *16* lines which read address **08h** of the EEPROM, add 5 to this value, and then store the result in EEPROM address **09h**. Finally, the program should loop until the write operation has finished.

A final point to note is that if you are not using interrupts at all (and have therefore disabled the global interrupt bit at the beginning of the program) you may remove the relevant lines in the EEPROM write procedure.

Example project: 'Telephone card chip'

You may be familiar with the so-called 'smart cards' which have found their way into a variety of applications. These cards have tiny chips embedded inside them, and either have contacts to communicate with the outside world or have a loop antenna inside the card. To demonstrate the use of the EEPROM, we will write the program for a PIC microcontroller which has been embedded within a 'smart' telephone card. We shall assume the card has 8 contact pins with which to communicate with the public telephone box. Two of these will be power pins, one will indicate that a call is in process and another will be from the card to alert the phone box when the card runs out of minutes. There will also be a pin used to reset the number of minutes left on the card to a specified value (this could be used for top-ups), and three pins will hold the 3-bit 'top-up' value, therefore allowing one of 8 possible values to be written to the card. The time remaining (in minutes) will be stored in the EEPROM so that the data remains intact when power is removed (the card is removed from the phone box). For simplicity's sake minutes are used as the basic unit of time, but you can adjust the program to count down in seconds, if you want.

The circuit diagram for this arrangement is shown in Figure 4.5. The interface with the phone box can be simulated using a number of switches and LED. GP0

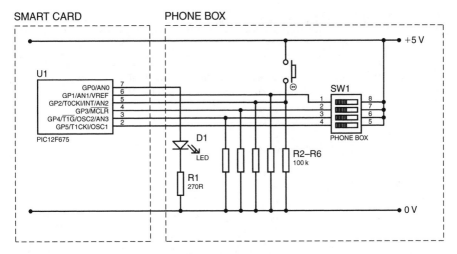

Figure 4.5

is an output and will go high if there are minutes on the card, and low when the time runs out. GP1 is an input from the phone box which goes high when a call is in progress. GP2 is an input which will be made high when the card value is to be reset, and is otherwise low. GP3, 4 and 5 store the 3-bit 'top-up' value. The flowchart is shown in Figure 4.6.

Starting from the program template developed earlier, we have to select values for INTCON, TRISIO, WPU and OPTION_REG. We will use the INT interrupt on GP2 (*rising* edge) to trigger the reset of the time remaining, and require no other interrupts. We will not use the weak pull-ups.

Exercise 4.8 What numbers (in binary) should be moved into INTCON, TRISIO, WPU and OPTION_REG in this project? Implement these changes in the program template.

From the flowchart, at **Main** we should first read the EEPROM and see if there are any minutes left on the card. The number of minutes will be stored in EEPROM address **00h**. If there are minutes left, the program should skip forward to a section called **Active**, and if not, it should turn off GP0 and go to sleep. If the card is topped up, the INT interrupt will trigger, the line following **sleep** will be executed and then the ISR will be called. After returning from this, the program should loop back to **Main**.

Exercise 4.9 **Challenge!** What *11* lines make up this first section? Don't forget to move back into Bank 0 as soon as you are able to. (**Hint:** The instruction **movfw** triggers the zero flag, if zero is moved into the working register.)

In **Active**, you should set GP0, and then enter a loop where you wait until there is a call in progress. When the call is active, the program should count time and

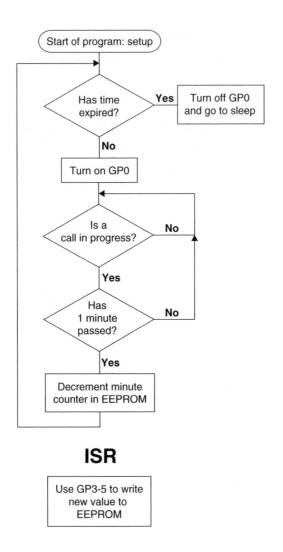

Figure 4.6

see if one minute has passed. Because the intermediate timing registers (markers and postscalers) are not stored in the EEPROM, the card will only count down complete minutes, and not fractions of minutes. This could easily be rectified by storing these registers in the EEPROM as well, but this is left as a possible development. Using the internal 4 MHz oscillator and TMR0 prescaled by 256, we can use a marker of 125 and postscalers of 125 and 15 to time one minute.

Exercise 4.10 Write the *15* lines which test to see if a call is in progress, and continue looping until one minute has passed.

Don't forget to set up the timing registers in **Init**. After one minute has passed we should reset the final postscaler to its correct value, and then decrement the number of minutes stored in the EEPROM. This involves reading the data in, decrementing it, and then writing it back. Finally, the program should loop back to **Main**.

Exercise 4.11 Write the *18* lines which complete this final section. Don't forget to clear the global interrupt enable before initiating the EEPROM write, and remember to set it again afterwards. The program must wait for the write operation to finish before looping back to **Main**.

All that remains is the handling of the INT interrupt, upon which the number of minutes stored should be reset to the value determined by bits GP3:5. The eight possible values are 2, 5, 10, 20, 40, 60, 120 and 0 minutes – assigned to (000, 001 . . ., 111) respectively. We should begin by clearing the INT interrupt flag (INTCON, 1), and then read in the state of GPIO. Don't forget that the interrupt may have occurred anywhere in the program, so we need to make sure we switch into Bank 0 in order to read the GPIO register. The bits of interest in GPIO are bits 3:5, so we should rotate this three times to the right. This can't be done directly to GPIO, so we have to use an intermediate register called **temp**. Finally, to turn this into a number between 0 and 8 (b'000' and b'111') we should mask bits 3–5.

Exercise 4.12 What *eight* lines clear the INT flag and then use bits 3–5 of GPIO to generate a number between 0 and 8, which is left in the working register?

We can create a lookup table in a subroutine called **CardValue** which is called with a number between 0 and 8 in the working register, and which returns with the appropriate number of minutes.

Exercise 4.13 Write the lookup table **CardValue**.

After calling **CardValue**, the number in the working register should be written into the EEPROM – you should be well practiced at this by now. Finally, you should wait until the write operation has completed before restoring the original values of STATUS and the working register. Upon returning, global interrupts should be enabled, so the **retfie** instruction should be chosen.

All that remains is a quick check to make sure you have declared all file registers, and set them up with appropriate values in **Init**. The entire program is shown in Program O. When simulating this project in MPLab, you can view the contents of the EEPROM by going to Yiew → EEPROM. When you program a PIC microcontroller, the contents of the EEPROM window will be written to the EEPROM on the chip. Similarly, when you read the program from a chip (Programmer → Read device), the contents of the chip's EEPROM will be shown in the EEPROM window.

Table 4.1

Middle C	C#	D	D#	E	F	F#	G	G#	A	A#	B
262 Hz	277	294	311	330	349	370	392	415	440	466	494

Further EEPROM examples: Music maker

For further EEPROM practice, you could make a device on which you can store musical notes and play back a melody. You can cover 8 octaves which correspond to 96 different notes, so each note is assigned a byte in the EEPROM. With an EEPROM of 128 bytes, a melody of up to 128 notes can be stored. A speaker can be connected to one of the outputs, and the different notes are obtained by producing square waves of different frequencies. The frequencies of the notes in the scale are shown in Table 4.1.

The notes of the other octaves are produced by multiplying or dividing these numbers by two. For example, the next C above middle C would be 524 Hz. Human hearing goes from about 10 Hz to 20 kHz, so rather than storing the frequency of the note in the EEPROM, it would be more sensible to store the note (e.g. D# or G) and the octave number (i.e. a number between 1 and 8). Each of these would be stored in a nibble, so for example, the hexadecimal number **56h** in the EEPROM could mean an E in the 6th octave.

Power monitor

A second possible EEPROM project is a device which is powered by the mains (indirectly of course!) and which counts and displays the time, storing the latest values in the EEPROM. Then, if the mains cuts out, when the device is next powered up it will display the time stored in the EEPROM – i.e. the time at which the power cut began.

Other applications for the EEPROM include devices concerning security where passwords are involved.

Analogue to digital conversion

Analogue to digital conversion (ADC) is the ability to measure the voltage at an analogue input and convert this reading into a number between 0 and 1024 (for 10-bit conversion). This translates into a precision of about 5 mV when a 5 V supply is used. For example, if the result of an A/D conversion was 10, the input voltage was 0.05 V, and if it was 400, the input was about 1.95 V. This allows much greater flexibility than digital inputs which can only tell whether an input is high or low (more than 2.5 V or less than 2.5 V). Some PIC microcontrollers support 8-bit ADC, which leads to lower precision. The voltage can be measured relative to the supply voltage (V_{DD}), or relative to the voltage on another pin (the V_{REF} pin – GP1).

A/D conversion can be a fairly lengthy process (compared with the speed at which most instructions are executed). The time which an A/D conversion takes can be changed by you, though if you make it too short the accuracy of the result will be affected. This and other aspects of the ADC are controlled in the registers **ADCON0** and **ANSEL**.

ADCON0

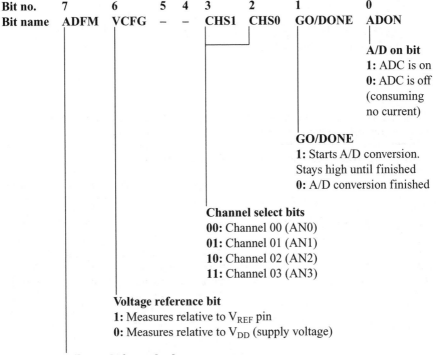

Bit no.	7	6	5	4	3	2	1	0
Bit name	ADFM	VCFG	–	–	CHS1	CHS0	GO/DONE	ADON

A/D on bit
1: ADC is on
0: ADC is off (consuming no current)

GO/DONE
1: Starts A/D conversion. Stays high until finished
0: A/D conversion finished

Channel select bits
00: Channel 00 (AN0)
01: Channel 01 (AN1)
10: Channel 02 (AN2)
11: Channel 03 (AN3)

Voltage reference bit
1: Measures relative to V_{REF} pin
0: Measures relative to V_{DD} (supply voltage)

A/D result formed select
1: Right justified – result stored in ADRESL and ADRESH (bits 0:2)
0: Left justified – result stored in ADRESL (bits 6:7) and ADRESH

Bit 0 of ADCON0 is the on/off switch for the A/D converter. When it is set, the ADC is on and the PIC microcontroller consumes extra current. Bit 1 is set to start an A/D conversion, and stays set for the duration of the process, after which it automatically clears. This bit can therefore be tested to see when the A/D conversion finishes. Bits 2 and 3 together select which analogue input you want to measure. For example, to test the voltage on AN2 (GP2) you should set bit 3 and clear bit 2. The voltage reference (V_{REF} pin or V_{DD}) can be selected using bit 6 of ADCON0. The measured 10-bit answer is held over two registers: **ADRESH** and **ADRESL** (**A/D Res**ult, **h**igher and **l**ower bytes). You have a choice in how the 10-bit number is stored over these two registers. Either it can be shifted to the right, so that bits 0:7 of the answer are held in ADRESL and bits 8:9 of the answer stored in bits 0:1 of ADRESH, or it can be shifted to the left, so that bits 0:1 of the answer are held in bits 6:7 of ADRESL, and bits 2:9 of the answer are held in ADRESH. This is illustrated in Figure 4.7.

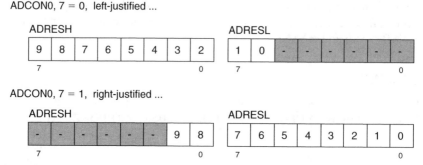

Figure 4.7

Table 4.2 Use of the ANSEL bits 2:0 to select the A/D conversion time. *Italic numbers represent conversion times which are too fast, or needlessly slow*

ANSEL bits 6:4	A/D conversion clock	Device frequency			
		1.25 MHz	**2.46 MHz**	**4 MHz**	**20 MHz**
000	Fosc/2	1.6 μs	*800 ns*	*500 ns*	*100 ns*
001	Fosc/8	6.4 μs	3.2 μs	2 μs	*400 ns*
010	Fosc/32	*25.6 μs*	*12.8 μs*	8 μs	1.6 μs
011	FRC: Internal oscillator	~4 μs	~4 μs	~4 μs	~4 μs
100	Fosc/4	3.2 μs	1.6 μs	*1 μs*	*200 ns*
101	Fosc/16	*12.8 μs*	6.4 μs	4 μs	*800 ns*
110	Fosc/64	*51.2 μs*	*25.6 μs*	*16 μs*	3.2 μs
111	FRC: Internal oscillator	~4 μs	~4 μs	~4 μs	~4 μs

ANSEL: Analogue select register

The ANSEL register has two purposes: setting the A/D conversion speed and selecting whether particular GPIO pins should be acting as analogue inputs, or standard digital I/O pins. Bits 0:3 refer to pins AN0:AN3 – when they are clear the relevant pin behaves like a digital I/O pin, however when set, the corresponding pin acts as an analogue input, and cannot be used as a digital input.

Example 4.7 Push buttons are connected to pin GP0 (AN0) and GP2 (AN2), while a thermometer input (analogue input) is connected to GP1 (AN1) and a microphone input (analogue input) to GP4 (AN3). The number **b'1010'** should be moved into bits 0:3 of ANSEL.

Bits 4:6 of ANSEL determine the A/D conversion clock, as shown in Table 4.2. Accurate A/D conversion requires a time of 1.6 μs or greater, however there is no point in making it much longer than this. The internal oscillator provides a conversion time of about 4 μs, though this can vary between 2 and 6 μs.

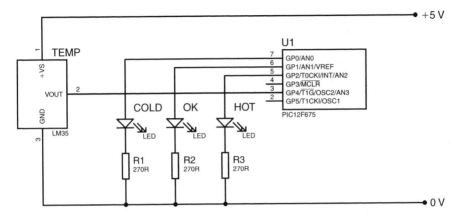

Figure 4.8

A/D conversion interrupt

To wait for an A/D conversion to complete, we could just keep testing ADCON0, bit 1 (which we used to start the conversion) and wait for it to clear. The A/D conversion interrupt frees up the program from this loop, and triggers upon completion of the conversion. This interrupt is a 'peripheral interrupt', and so it is enabled in the PIE1 register (bit 6) and its interrupt flag is found in the PIR1 register (bit 6). In order for the interrupt to trigger, both the peripheral interrupt enable and the global interrupt enable bits in INTCON (bits 6 and 7) must be set.

Example project: 'Bath monitor'

To practise A/D conversion, our next project will be a temperature-sensing device which indicates whether the temperature of your bath is too high, too low, or just right (i.e. within an acceptable temperature range). There will be three LEDs to indicate these three possible conditions, connected to GP0, GP1 and GP2. GP4 (AN3) will be the analogue input connected to the temperature sensor LM35 which varies its output linearly according to temperature. The circuit diagram is shown in Figure 4.8, and the flowchart in Figure 4.9.

As with the quiz game controller, the main loop of the program is practically nothing at all. In this case we simply need to keep starting the A/D conversions, and the response to the measurement will be handled in the ISR. The program from **Start** is therefore:

```
Start    call   Init        ; sets everything up
Main     bsf    ADCON0, 1   ; start A/D conversion
         goto   Main        ;
```

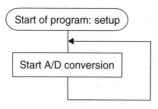

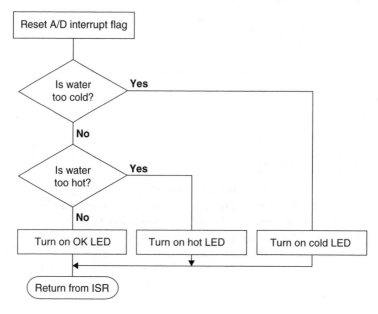

Figure 4.9

Looking at the above code we can see that using an A/D conversion interrupt in this case isn't actually necessary. We could simply have written the following:

```
Start    call    Init        ; sets everything up
Main     bsf     ADCON0, 1   ; starts A/D conversion
ADLoop   btfsc   ADCON0, 1   ; has conversion finished?
         goto    ADLoop      ; no, so keeps looping
         etc.                ; yes, so exits loop
```

However, in a more advanced version of this program we may want to have a more complex main loop, and in such a case the A/D conversion interrupt may be very useful. We will therefore keep using the interrupt method in this project and write the program as if the interrupt may have occurred during a more complex program (i.e. use the working and STATUS register storage/recovery code in the **isr**).

Exercise 4.14 Specify the numbers that should be moved into the following registers during the **Init** subroutine: INTCON, ADCON0, ANSEL, TRISIO, WPU, OPTION_REG and PIE1. Assume the internal 4 MHz oscillator will be used, and set the result of an A/D conversion to be *left-justified*.

In the ISR, we needn't test the A/D interrupt flag as it is the only interrupt which has been enabled, but we do need to reset it (clear it).

Exercise 4.15 What *two* lines will reset the A/D interrupt flag (make no assumption about the current bank)?

Following the flowchart, we see that the next step is to see whether the temperature is too cold (i.e. whether the measured analogue voltage is below a certain threshold). Depending on the required accuracy, we have two choices. We can choose to discard the two *least significant* bits of the answer (bits 0 and 1) which are held in ADRESL, and use only the 8 bits held in ADRESH. We can simply see whether ADRESH is below the 'cold' threshold. This particular temperature sensor gives an output voltage of 0.01 V per degree Celsius. If we say the minimum bath water temperature is 36°C, this means the minimum input voltage is 0.36 V (which is compared with the reference voltage, V_{DD}, which is 5 V). $0.36/5 = 0.072$ and $0.072 \times 256 = 18$. We've multiplied by 256 because by using only the number in ADRESH (the *eight* most significant bits). We would therefore write:

```
movlw   d'18'        ; is ADRESH less than 18?
subwf   ADRESH, w    ;
btfss   STATUS, C    ; carry flag is 0 when result is -ve
goto    Cold         ; C is 0, therefore ADRESH < 18
```

If, on the other hand, we want to take advantage of the full 10-bit precision of the A/D converter, we need to test the full 10-bit number which is spread over ADRESH and ADRESL. For the sake of the more discerning bather, we will take this 10-bit approach in this project. The threshold voltage for 36°C is still 0.36 V, which when divided by 5 V leaves 0.072. We multiply this by 1024 to get the 10-bit value for the cold threshold: **d'74'**. If you write this number out in binary as it will appear in registers ADRESH and ADRESL you get: **b'00010010 10000000'** (remember – bits 0:1 of the A/D result are stored in ADRESL bits 6:7, and bits 2:9 of the A/D result are stored in ADRESH). Thus, the upper byte is (**0x12**) and the lower byte is (**0x80**). To test the result of the 10-bit A/D conversion we subtract a number from the result, as we have done before, except this time the number happens to be split over two file registers. We handle this in the same we handle normal arithmetic – first subtract the lower bytes, subtracting one from the higher byte if you need to borrow, then subtract the higher bytes.

```
bsf     STATUS, RP0  ; goes to Bank 1
movlw   0x80         ; subtracts lower byte
subwf   ADRESL, w    ;
```

```
comf    STATUS, w       ; inverts carry flag (bit 0 of STATUS)
andlw   b'00000001'     ; masks all other bits
bcf     STATUS, RP0     ; goes to Bank 0
addlw   0x12            ; add this to the number we are
subwf   ADRESH, w       ;  subtracting from the higher byte
btfss   STATUS, C       ;
goto    Cold            ; ADRESH:L < 0x1280, → "cold!"
etc . . .
```

We first subtract the lower byte of the threshold from the lower byte of the answer (ADRESL), leaving ADRESL unaffected. Note that we don't care what the answer is – only whether we had to borrow or not. If we borrow, the carry flag is clear. We invert all the bits of STATUS (including the carry flag), moving the result into the working register, then mask all bits other than bit 0 (which is now the carry flag – *inverted*). This means the working register is now 0 if there was no borrow, and 1 if there was a borrow. We can therefore add the working register (0 or 1) to the number we want to subtract from the higher byte, then subtract the total from ADRESH. Again, we're not interested in the answer itself, only in how the carry flag was affected. If it's clear, there was a borrow, meaning that the overall number split over ADRESH and ADRESL was less than **0x1280**, and the bath is therefore cold.

Exercise 4.16 The maximum temperature shall be 42°C. How does this value translate into the 10-bit number produced by the A/D converter, and how is it distributed over ADRESH and ADRESL.

Exercise 4.17 Write the *11* lines which use the technique described above to test to see if the temperature is too hot. If it is too hot, branch to a section called **Hot**. If it's not too hot, we've already tested whether it's too cold, so we know it's OK and so branch to a section called **OK**.

Each section (**Cold**, **OK** and **Hot**) should turn the correct LED on, and then return while enabling the global interrupt enable. Rather than copying out the code for restoring the values of the working register and STATUS, you can label this section **prereturn**, and jump to **prereturn** at the end of the three different sections.

Exercise 4.18 Write the **Cold**, **OK** and **Hot** sections. They should consist of three lines each.

The entire program is now complete, and shown in Program P. If you are using the *PICKit™ 1 Flash Starter Kit*, you can use the components already on the board to test the program, though you will have to change some of the pin assignments. You should change the analogue input to GP0, and you can then simulate different temperatures by turning the potentiometer. The LEDs can be controlled on by making pins GP1, 2, 4 and 5 inputs, outputs = 0 or outputs = 1, as required. For example, to turn on LED0, make TRISIO = **b'001111'** and GPIO = **b'010000'**; to turn on LED1, make TRISIO = **b'001111'** and GPIO = **b'100000'**; and to turn on LED2, make TRISIO = **b'101011'** and GPIO = **b'010000'**.

Comparator module

On the surface, the comparator module looks like a simplified version of analogue to digital conversion. A comparator measures two analogue inputs, called V_{IN+} and V_{IN-}, and produces a digital output V_{OUT} depending on which voltage is bigger. In the standard configuration, $V_{OUT} = 1$ if $V_{IN+} > V_{IN-}$ and $V_{OUT} = 0$ if $V_{IN+} < V_{IN-}$, however, this behaviour can be inverted when configuring the comparator.

Within this fairly basic type of operation, the PIC microcontroller offers a wide range of different possible forms of behaviour which are controlled by bits 2:0 of **CMCON** (**C**omparator **M**odule **Con**trol register), and summarised in Table 4.3. On the PIC12F675, comparator inputs can be chosen from $GP0/C_{IN+}$, $GP1/C_{IN-}$ or even a programmable internal voltage reference. V_{OUT} can be directly connected to pin $GP2/C_{OUT}$, or else can be released and used as a standard digital I/O pin. In the latter case, the comparator output can be read by the program as bit 6 of CMCON.

If GP0 is not being used by the comparator (e.g. type '**B**' behaviour), it can be used as a standard digital I/O pin. In the case where either GP0 or GP1 can be used as the V_{IN+} input (i.e. type '**C**'), *both* are set as analogue inputs. Note that when the PIC microcontroller is powered up or reset, CMCON 2:0 is 000, and even though the comparator is disabled (V_{OUT} is set to 0), pins GP0 and GP1 remain analogue inputs and cannot be used as digital inputs. Hence, if you are not using the comparator, it should be turned off by setting CMCON 2:0 to 111. As is the case for any analogue inputs, the voltage must be within the supply voltages V_{SS} and V_{DD}.

We've already discussed bits 0:2 and bit 6 of CMCON, and we will now examine the remaining bits. In type '**C**' behaviour, bit 3 is the Comparator Input

Table 4.3

CMCON 2:0		V_{IN+}	V_{IN-}	V_{OUT}
000		$GP0/C_{IN+}$	$GP1/C_{IN-}$	*Disabled: CMCON, 6 = 0*
001	**A**	$GP0/C_{IN+}$	$GP1/C_{IN-}$	$GP2/C_{OUT}$ *and* CMCON, 6
010		$GP0/C_{IN+}$	$GP1/C_{IN-}$	CMCON, 6
011	**B**	Internal ref.	$GP1/C_{IN-}$	$GP2/C_{OUT}$ *and* CMCON, 6
100		Internal ref.	$GP1/C_{IN-}$	CMCON, 6
101	**C**	Internal ref.	GP0 *or* GP1	$GP2/C_{OUT}$ *and* CMCON, 6
110		Internal ref.	GP0 *or* GP1	CMCON, 6
111		*Comparator off and consumes no current (CMCON, 6 = 0)*		

Table 4.4

VRCON, 5 = 1 (low range)		VRCON, 5 = 0 (high range)	
VRCON 3:0	**VRef (V_{DD} = 5 V)**	**VRCON 3:0**	**VRef (V_{DD} = 5 V)**
0000	0.00	0000	1.25
0001	0.21	0001	1.41
0010	0.42	0010	1.56
0011	0.63	0011	1.72
0100	0.83	0100	1.88
0101	1.04	0101	2.03
0110	1.25	0110	2.19
0111	1.46	0111	2.34
1000	1.67	1000	2.50
1001	1.88	1001	2.66
1010	2.08	1010	2.81
1011	2.29	1011	2.97
1100	2.50	1100	3.13
1101	2.71	1101	3.28
1110	2.92	1110	3.44
1111	3.13	1111	3.59

Switch which selects whether GP0 or GP1 is being measured. Finally, bit 4 is the Comparator Output Inversion bit – when this is set, any output from the comparator is inverted.

Voltage reference

As well as comparing the states of external analogue inputs, the comparator can use an internal programmable voltage reference. In order to use it, we must first turn on the voltage reference module by setting bit 7 of the **VRCON** register (**V**oltage **R**eference **Con**trol). The voltage reference can take one of 32 distinct values, as given by bits 5 and 3:0 of VRCON. Bit 5 selects one of two voltage ranges; when set, the lower range of voltages is selected, and the reference is equal to V_{DD} * (VRCON 3:0)/24. When bit 5 is clear, the upper range is selected, and the reference is equal to V_{DD}/4 + V_{DD} * (VRCON 3:0)/32. Table 4.4 above shows example values for voltage references, for V_{DD} = 5 V.

As with the comparator module, don't forget to turn off the voltage reference to save power if you aren't using it, or when going into sleep mode. On some PIC microcontrollers (e.g. the PIC16F627), this reference voltage can be output through an I/O pin.

Comparator interrupts

The comparator interrupt triggers when the state of the comparator output changes. The corresponding interrupt enable is found in the PIE1 register (bit 3), and the interrupt flag is stored in the PIR1 register (bit 3) – this must be reset to 0 after the interrupt occurs. As with all peripheral interrupts, both the PIE Enable and Global Interrupt Enable bits must be set for this interrupt to occur. If you plan to change the comparator behaviour during the program, you should disable the comparator interrupt during the change, to avoid the possibility of a false interrupt.

Comparator example: 'Sun follower'

As a short example of the comparator feature, we will look at the program for a solar cell 'sun follower'. There are two sensors, on either side of the solar cell, which measure the light level and produce analogue voltages between 0 and 5 V. As the sun rises and sets during the day, we want the solar cell to point directly towards the sunlight. The device therefore compares the signal coming from each sensor, and then drives a motor to make the two equal. The two light sensors are connected to the GP0/C_{IN+} and GP1/C_{IN-} pins. The motor is connected to pins GP2 and GP4; if GP2 is high and GP4 is low, the motor is driven forward, if GP4 is high and GP2 is low, the motor is driven backward. Every ten minutes a subroutine will be run to adjust the solar cell position. We begin by turning on the comparator module, and making the correct settings. We wish to compare GP0 and GP1 (i.e. type **'A'** operation) and do not require the C_{OUT} pin, hence CMCON 2:0 should be **b'010'**. We aren't using type **'C'** operation so bit 3 doesn't matter, and we don't wish to invert the comparator output so bit 4 should be 0:

```
FollowSun   bcf     STATUS, RP0     ; Bank 0
            movlw   b'00000010'     ; turns on comparator, compares
            movwf   CMCON           ;  GP0 & GP1, not using COUT pin
```

The comparator response time can be as long as 10 µs when a new input or voltage reference has been chosen, or when just turned on, so we need to insert a short delay before responding to the comparator output. An easy way to do this is to create a subroutine, **delay**, which immediately returns:

```
delay       return                  ; immediately returns
```

Calling this subroutine and returning will take four clock cycles (or 4 µs, given the internal 4 MHz internal oscillator). Depending on the comparator output, the motor is driven forward or in reverse:

```
            call    delay           ; kills four clock cycles (4 µs)
            call    delay           ; kills four clock cycles (4 µs)
            call    delay           ; kills four clock cycles (4 µs)
            bcf     PIR1, 3         ; resets comparator interrupt flag
            btfss   CMCON, 6        ; reads comparator output
            goto    Forward         ; drives motor forward
Reverse     bsf     GPIO, 4         ; drives motor in reverse
```

```
            goto    Continue    ;
Forward     bsf     GPIO, 2     ; drives motor forward
Continue    ...
```

We then wait for a comparator interrupt (we don't actually need to use an interrupt – it's easier to simply test the comparator interrupt flag, which will get set even if the relevant interrupt enable bits are disabled). The comparator interrupt will take place when the comparator output changes – i.e. just when the values from the light sensors are approximately equal. At this point we can stop the motors, and then return from the subroutine. This assumes the motors are sufficiently slow that overshoot isn't a problem.

```
Continue    btfss   PIR1, 3     ; waits for comparator to change
                                ;  output

            goto    Continue    ;
            bcf     GPIO, 2     ; turns off motor
            bcf     GPIO, 4     ;
            return              ; returns from 'FollowSun'
                                ;  subroutine
```

As well as showing how the comparator may be used, the above example illustrates how an interrupt flag may be used without actually involving a jump to the interrupt service routine.

Comparator example: Reading many buttons from one pin

We can also use the comparator to read a large number of buttons from only one input. If we connect the buttons as shown in Figure 4.10, there is a different resistance between the GP1/C_{IN-} pin and V_{DD}, depending on which button is pressed. We also place a capacitor between GP1 and ground, so that when a button is pressed there is a slow rise time which is dictated by the values of resistance and capacitance. Therefore, the rise time is different for each button. By using the comparator, we can set a particular threshold voltage for the GP1 input. We discharge the capacitor by making GP1 an output and setting it to 0, then we make it an input and set a timer going. By measuring the time at which the comparator output changes, we can identify which button was pressed (if any).

With the values given in Figure 4.10, the discharge is very fast. The value of $R \times C$ for each button is: 8 μs, 23 μs, 38 μs or 53 μs, and is 68 μs if there is no button pressed. If we set up the comparator to trigger at about $(1 - 1/e) \times V_{DD} = 3.16$ V, the trigger times should be equal to the RC products given above. The code below could therefore be used to read the buttons:

```
Setup   bsf     STATUS, RP0     ; Bank 1
        movlw   b'10001000'     ; TMR0 not prescaled (counts up
        movwf   OPTION_REG      ;  once every 1 μs for 4 MHz clock)
        movlw   b'10001100'     ; programs an internal voltage
```

```
                movwf   VRCON           ;  reference of 3.13 V
                bcf     STATUS, RP0     ; Bank 0
                movlw   b'00000100'     ; turns on comparator, to compare
                movwf   CMCON           ;  GP1 with internal V_REF
ButtonTest      bsf     STATUS, RP0     ; Bank 1
                bcf     TRISIO, 1       ; make GP1 output
                bcf     STATUS, RP0     ; Bank 0
                bcf     GPIO, 1         ; discharge capacitor
                clrf    TMR0            ; resets TMR0
                bsf     STATUS, RP0     ; Bank 1
                bsf     TRISIO, 1       ; make GP1 input ← TMR0 = 0
                bcf     STATUS, RP0     ; Bank 0
Loop            btfsc   CMCON, 6        ; waits until V_IN− > V_REF
                goto    Loop            ;
                swapf   TMR0, w         ; takes current value of TMR0
                andlw   b'00000111'     ; takes only bits 4:6 of TMR0
                addwf   PCL, f          ; skips between 0 and 4
                goto    Button1         ;  instructions, depending on
                goto    Button2         ;  which button, if any, was
                goto    Button3         ;  pressed
                goto    Button4         ;
                goto    NoButton        ; no button was pressed
```

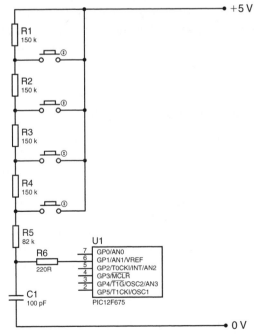

Figure 4.10

In the setup section we choose no prescaler for TMR0 (so, given a 4 MHz clock, it counts up every 1 μs). We choose an internal voltage reference of 3.13 V, and turn on the comparator, selecting $GP1/C_{IN-}$ and V_{REF} as the comparator inputs, and using only CMCON, 6 as the comparator output. During the button test we first discharge the capacitor, then reset TMR0. Note that when you write to TMR0 (e.g. clear it), the change takes *two cycles* to take effect. Therefore TMR0 isn't actually cleared until the line in which GP1 is made an input, which is just what we want. We then enter a loop which waits until the comparator input goes high (the voltage has risen above the threshold). We then take the number from TMR0 and use it to jump to the appropriate section.

You will now notice that the resistor values were not chosen at random – if Button1 was pressed, the expected rise time is 8 μs. TMR0 counts up once every 1 μs, so the expected value of TMR0 is 8. Even if there is a small error, we can be pretty certain the number in TMR0 is **b'0000????'**, i.e. the value of TMR0 bits 4:6 should be **000**. If Button2 was pressed, the expected rise time is 32 μs, which means TMR0 bits 4:6 are expected to be **001** (work it out!). So, our choice of resistor values means that even with some small timing errors, the values of TMR0 bits 4:6 will be: 000, 001, 010, 011 or 100 depending on which button (if any) has been pressed. Therefore, when the threshold voltage is reached, we *swap* the nibbles in TMR0 (making the bits of interest bits 0:2), leaving the result in the working register, and then mask the answer using the AND operation. This provides a number between 0 and 4 which can be added to the PC to branch to the different sections. The whole read operation therefore takes less than 100 μs, though an important drawback of this method is the inability to detect when more than one button is pressed. Finally, in practice, you should check carefully the real values of the resistors and capacitor you are using, and be prepared to play with the voltage reference value to achieve the desired behaviour.

Final project: Intelligent garden lights[1]

We will bring together some of the ideas covered in this chapter in a final project: an intelligent garden lights unit (thanks to Max Horsey for the idea and original design). This device detects the ambient light level, and according to user programming, turns on the garden lights when it gets dark. The lights are automatically turned off around midnight. The user-programmed settings will be stored in the EEPROM, in case of loss of power. There will be an override button which allows the lights to be manually turned on and off, and a switch which is used to tell the device whether we are currently on 'daylight savings time'. The key behind this project is the rule-of-thumb that midnight roughly coincides, within

---

[1] In previous editions of this book, the final project was a Lottery Number generator with 'vibe' detection, giving personalised numbers and special messages. Due to the popularity of this project, it is described on the supporting website: www.to-pic.com.

20 minutes or so, with the 'solar midnight' – the halfway point between sunrise and sunset. In other words, the midpoint between the time for a particular light level in the evening and time for *the same* light level in the early morning is approximately midnight. For example, if it gets dark around 7 p.m., the light level should be approximately the same at 5 a.m. (so that the midpoint between these times is 12 a.m.). This means the device can calculate midnight without the need for the user to input a time (and the clumsy interface this may entail).

The override button will trigger an external INT interrupt and so will be connected to the GP2/INT pin. The garden lights will be controlled through GP5 (using a relay in the real device, or simply an LED in the test version). GP4 will control whether the 'day' or 'night' LED is on – which tells the user whether the device thinks it is currently day or night. The light sensor will be attached to GP0/AN0, and the summer/winter switch to GP1. Finally, when the PIC microcontroller is reset (using the GP3/$\overline{\text{MCLR}}$ pin) it will measure the current light level and use this as the threshold light level at which to turn on the garden lights. A button attached to this pin is therefore pressed to program the light level at which garden lights should be turned on. The flowchart for this project is shown in Figure 4.11, and the circuit diagram in Figure 4.12.

We'll go through the key steps of the flowchart and the program, but the actual writing of the program is left as an exercise. The program I wrote is shown in Program Q, but yours may differ in parts. First, the $\overline{\text{POR}}$ bit in the **PCON** (**P**ower **Con**trol) register is used to determine whether the device has just powered up, or been reset by the $\overline{\text{MCLR}}$ pin. If the device has just been powered up, the $\overline{\text{POR}}$ bit will be 0 (and needs to be reset to 1), and the last-saved values can be read out from the EEPROM. The 'day' LED is turned on and the program waits for the light to fall below the threshold (i.e. wait for dusk). However, if a reset occurred, the light level is measured and stored as the new threshold. The value for midnight is also reset. In this application, 10-bit accuracy is not required from the A/D converter, so the 8-bits in ADRESH can be used (given a left-justified A/D result).

Once the garden lights come on, a timer is started both to determine when to turn off the lights and also to measure when tomorrow's midnight will be. You can time in units of five minutes (using a marker of 125, and postscalers of 125 and 75). Five minute accuracy is sufficient, and it allows you to time a whole night using one file register (it times a maximum of 5 mins $\times$ 256 = over 21 hours, which should be enough for most places, except perhaps a winter in Lapland!). Given that light levels may fluctuate slightly, we won't do anything at all for the first hour. After this point, we wait until midnight, in other words, we wait until time elapsed equals the previously estimated value for the time between dusk and midnight. We then turn off the garden lights (leaving the 'night' LED on).

Finally, we test the light levels to wait for dawn (whilst keeping the timing going). When the light levels exceed the threshold, we store the elapsed time (the time from dusk until dawn) and divide it by two. This gives our estimate for the time between dusk and midnight. If we are in summer time (the clocks have gone forward one hour), our guess is out by two hours and so we should subtract two hours from the estimate. This value should then be stored in the EEPROM,

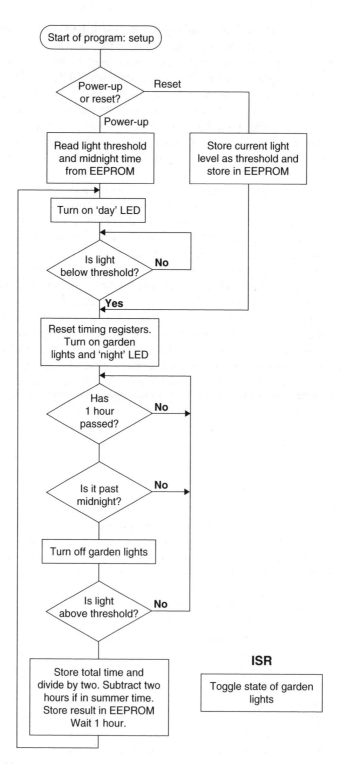

Figure 4.11

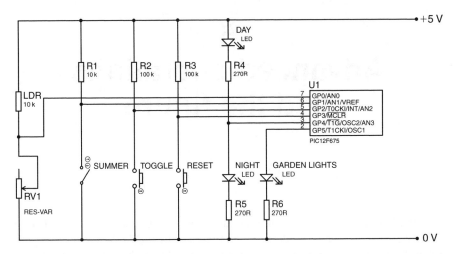

Figure 4.12

and the program should loop back. We should wait one hour before testing for dusk again, to minimise the risk of errors. The override button will trigger the INT interrupt, which simply has to toggle the state of GP5 to turn the garden lights on and off. This will not affect the normal operation of the device.

This program has combined the use of interrupts, A/D conversion and the EEPROM, and provided you with the opportunity to tackle a program on your own, with only basic guidance. You should now have the confidence and the tools with which to design and build your own PIC projects.

5
Advanced operations and the future

The market of programmable microcontrollers is doubtless one of the fastest growing areas in electronic design, and there are new PIC devices coming out all the time. Other microcontrollers (in addition to Microchip's PIC) are flooding the market, each with their own competitive edge, and all fighting for a piece of the action. The challenge from the user's point of view is keeping up with all these newcomers. As far as new PIC models are concerned, they will maintain the same basic structure, but with new features added here and there. For example, there may be a new special function with accompanying file registers, extra I/O pins, more timers, etc. The key to keeping up with these is recognising what you want, and learning how to interpret the accompanying datasheets. At first sight, these enormous manuals may seem undecipherable, but there are certain pages to look out for when trying to find out about a new feature. Such pages include the ones showing the banks of PIC file registers, which will allow you to spot any new ones. You can then use the index to find the relevant pages and learn more about the new features.

To aid in this endeavour, we will now briefly examine the kinds of advanced functions available on more complex PIC microcontrollers. Their detailed operations are beyond the scope of this book, but it is useful to know what may be available.

Extra timers: TMR1 & . . .

On some PIC models you may find a second timer called TMR1, which gives you the freedom to use one timer for counting signals, and one for timing, for example. TMR1 is a 16-bit timer, so its value is spread over two registers, **TMR1H** and **TMR1L**, which contain the higher and lower bytes respectively. Like the TMR0, it has an associated interrupt which triggers when the timer overflows (in this 16-bit case, going from FFFFh to 0000h). TMR1 is controlled by the **T1CON** register, which gives you the choice of counting from the internal oscillator, or an external signal on the T1CKI pin. The TMR1 can also be set to count only while the $\overline{T1G}$ pin is low. You will have to take care when reading the 16-bit number, because a byte of TMR1 might overflow over the course of your measurement.

Example 5.1 We wish to read the number in TMR1, which happens to be 28FFh.

TMR1Read

movfw	**TMR1H**	**; stores the number in the higher byte**
movwf	**TempH**	**; in the register TempH**
movfw	**TMR1L**	**; stores the number in the lower byte**
movwf	**TempL**	**; in the register TempL**

Let's say that halfway through the above code, TMR1 counts up to 2900h. The upper byte may have been read as 28h, and then the lower byte as 00h – leading to a substantial error in the read operation! To make a safe read, the above code should be followed by:

movfw	**TMR1H**	**; takes the current higher byte and**
subwf	**TempH, w**	**; compares it with stored value**
btfss	**STATUS, Z**	**; are they different?**
goto	**TMR1Read**	**; yes, so repeat measurement**
		; no – so no overflow: read is safe!

When writing to the TMR1, a similar problem may be encountered, but this can be avoided by simply stopping the TMR1 (using T1CON), writing the number, and then starting it again. There may also be further timers TMR2, etc. available.

Capture/Compare/PWM

On the PIC16F627, for example, there is a *Capture/Compare/PWM* module which can perform three distinct tasks. However, all tasks share the Capture/Compare/PWM Registers: **CCPR1H** and **CCPRIL**, and are controlled by **CCP1CON**. The *Capture* and *Compare* features integrate closely with the 16-bit TMR1, and the *PWM* feature uses a third timer, the 8-bit TMR2.

Let's say, for example, that we wish to measure the time until an event occurs on a certain pin. We could just test the pin, and then read the timer value, but in order to simplify the program and free up the processor on the chip, we can use the handy *capture* feature. *Capture* immediately stores the value in TMR1 (both the higher and lower bytes) when a certain event occurs. The value is automatically stored in registers CCPR1H and CCPR1L, which can be read in the standard way. Trigger events are limited to a particular pin (e.g. RB3/CCP1), but can take place on:

- Every falling edge
- Every rising edge
- Every 4th rising edge
- Every 16th rising edge.

The RB3/CCP1 pin must be configured as an input for this to work. A *capture* event can also trigger an interrupt.

In almost any application of timers/counters, you are testing to see if the timer has reached a certain value. *Compare* does this for you, in that it constantly

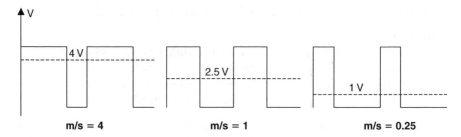

Figure 5.1

compares the number stored in CCPR1L/H with the number in TMR1. To wait until TMR1 reaches a particular value, you put the desired value into CCPR1L/H, enable *compare* mode, and then just wait. When there is a match, the RB3/CCP1 pin can be set to go high or low, or left alone, and an interrupt may be triggered.

PWM stands for Pulse Width Modulation, and refers to the ability to change the *mark-space* ratio in a square wave output (as shown in Figure 5.1). The mark-space ratio is the duration of the 'logic 1' part of the wave, divided by the duration of the 'logic 0' part. By controlling this ratio, we can control the output voltage, which is effectively an average of the square wave output (though you may need to add a resistor/capacitor arrangement depending on the application).

PWM can be used on the RB3/CCP1 pin (assuming it has been configured as an output) and has up to 10-bit resolution. Both the period of the square wave output, and the mark-space ratio can be controlled, and timing is performed by a separate timer called TMR2.

USART: Serial communication

USART stands for Universal Synchronous/Asynchronous Receiver/Transmitter, and allows the PIC microcontroller to communicate with a wide range of other devices from separate memory chips and LCD displays, to personal computers! This involves sending or receiving 8- or 9-bit packets of data (e.g. a byte, or a byte plus a parity bit). A parity bit is an extra bit sent along with the data that helps with the error checking. If there are an odd number of 1s in the data byte (e.g. b'00110100'), the parity bit will be 1, and if there are an even number (e.g. b'00110011'), the parity bit will be 0. In this way, if an error (e.g. a bit flip) occurs somewhere between sending the byte and receiving it, the parity bit will no longer match the data byte. The receiver will know that something has gone wrong, and it can ask for the byte to be resent. If *two* bit errors occur in one transmission, the parity bit will appear correct, however the probably of two errors occurring is substantially smaller, and so this is often overlooked.

The USART module has two principle modes: *asynchronous* and *synchronous* operation. In asynchronous operation, the transmitter pin (TX) from one device is connected to the receiver pin (RX) of another, and data is swapped (known as full-duplex). In synchronous operation, clock (CK) and data (DT) lines are shared

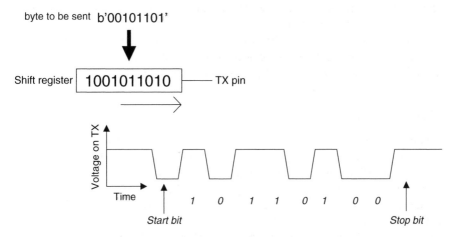

byte to be sent b'00101101'

Shift register | 1001011010 | —— TX pin

Voltage on TX

Time

Start bit   1 0 1 1 0 1 0 0   Stop bit

Figure 5.2

between a number of devices (one *master*, and one or more *slaves*). The master is responsible for producing the clock. In both cases, the rate at which data is sent by the transmitter (and at which it is expected by the receiver) is known as the *baud rate*.

There are two registers for controlling the receiving and transmission of data: **RCSTA** and **TXSTA**, respectively. Data that's successfully read is stored in **RCREG**, and data that's to be transmitted should be placed in **TXREG**. The baud rate is set using the **SPBRG** register (there are extensive tables in the datasheets showing how to select baud rates given certain oscillator frequencies, etc.).

In asynchronous mode, the USART takes the 8- or 9-bit character to be sent, and adds a *start bit* (a zero) to the front, and a *stop bit* (a one) to the end to create a 10- or 11-bit sequence. This is then moved onto a shift register which rotates the bits onto the transmission pin (TX), as shown in Figure 5.2.

The receiver module will constantly check the state of the RX pin, which will normally be high. If it detects the RX pin goes low (a potential *start bit*), it then makes three more samples in the middle of the bit (allowing for slow rise and fall times) and takes the majority value of the three. If the majority value is 0, it's convinced this is really a start bit, and carries on sampling subsequent bits with three samples in the middle of each bit. The timing of this sampling is dictated by the baud rate. When it reaches what should be the stop bit, it must read a one, otherwise it will declare the received character *badly framed* and register an error. Remember that with the appropriate settings in TXSTA and RCSTA, all this is done for you by the USART module.

You can use asynchronous mode to communicate with an RS232 serial port on your PC. A simple way to send bytes through your PC's serial port is through a program that comes with Microsoft® Windows® called HyperTerminal (Start Menu → Programs → Accessories → Communications). You can create a connection with your serial port (e.g. COM1), choose a baud rate, number of bits, parity setting, etc. When HyperTerminal connects to the serial port, whatever character you type is sent (as ASCII) through the serial port. Characters which are received are displayed on screen.

Both asynchronous and synchronous modes support a feature known as *address detect* which allow a number of devices to be connected. When transmitting data, an address byte must first be sent out to identify the intended recipient.

Programming tips

It is important that you don't jump into the deep end with program writing, and keep things simple to begin with. Furthermore don't sit down and try and write the whole thing all at once. Split the program up into key elements, and aim to get certain bits working as you go along. Simple things like periodic breaks and clear comments are important, and if you ever get stuck remember to ask someone, even if they've never seen a PIC microcontroller before.

The key to becoming a better programmer is very simple indeed – practice. All that it takes to be able to write programs efficiently and effectively is a bit of experience. Now that you have the knowledge to start writing your own programs you will find that you learn more and more. For example, as I was writing one of my first programs I came across something I had never realised before. I wanted to test to see whether or not TMR0 held the number zero, so I wrote the following:

movf   TMR0   ; is the number in TMR0 zero?
btfss   STATUS, Z ;

I found while simulating the program, that TMR0 just wouldn't count up. As I saw it, the PIC program was taking the number out of TMR0, and then putting it back in again. However, it then occurred to me that there needs to be something keeping track of exactly what the number in TMR0 is (e.g. 56 and three quarters). Although the integer part of the number (56) is held in TMR0, the fraction is held somewhere else. It became clear that whenever you move a number into TMR0, that fraction part gets cleared to zero. This explained why TMR0 was never getting anywhere, so I added that all important w:

movfw   TMR0   ; is the number in TMR0, zero?
etc.

As you can see, you never stop learning – and don't stop experimenting. The great thing about PIC microcontrollers is that you can try things out easily, and then forget it if it didn't work. Be sure to visit my PIC website at www.to-pic.com, where your PIC questions will be answered, and you will find helpful hints and the example programs in .txt format. So with this last piece of advice I leave you, good luck, and happy 'PICing'.

6
A PIC development environment

Some of you may feel daunted by loading a blank page in Notepad, or MPLab, and trying to begin writing your own program. There is a development environment dedicated to PIC programming for beginners, called 'PIC Press'.

First and foremost the software assembles each line of program as you write it, so you are instantly alerted if you have done anything wrong, and specific error messages are given, helping you spot the error quickly. The colour coding as you type instructions, labels, numbers, comments, and errors, make the program much easier to look at and interpret. Labels (i.e. things which you **go to** or **call**) come out in blue, however they are purple if broken – e.g. if you write **goto Main** and you haven't started the **Main** section, 'Main' will be purple to remind you it hasn't been started. Comments are made green, errors red, and instructions are made bold. A diagram of your file registers is provided showing you how your general purpose file registers are arranged, as well as reminding you how the special function registers occur for the particular PIC model you happen to be using.

When you first start up the software and begin a new program, you are asked questions regarding which PIC microcontroller you intend to use (offering help in selecting one), which clock frequency, which type of oscillator, what the name of the project is, etc. and then it automatically creates the header for the program, as well as making a note of the PIC model for use in other aspects of the software.

The software then asks you about the inputs and outputs of the PIC microcontroller, so that it can automatically fill in the **tris** aspect of the program, and then asks you how you would like to set up the special function registers for the PIC model you are using. Rather than having to look up the bit arrangements for each register for the particular PIC model you are using, all this is provided by the software in such a way that you can simply tick or clear boxes depending on whether you want that particular function turned on or off.

The software then creates the Init subroutine and fills in all the lines of code relating to the information which you have just provided. Now, rather than being faced with a blank screen, you are starting with the bones of your program in place, and left with filling in the flesh.

To help you further with writing the main body of the program, a selection of 'macros' are available with the program. For example there is one which automatically creates a time delay. You simply choose the length of time you wish to wait, and the software uses the clock frequency which you have already entered at the beginning, to create the code required. Other macros include 'EEPROM write', 'A/D conversion', and many others.

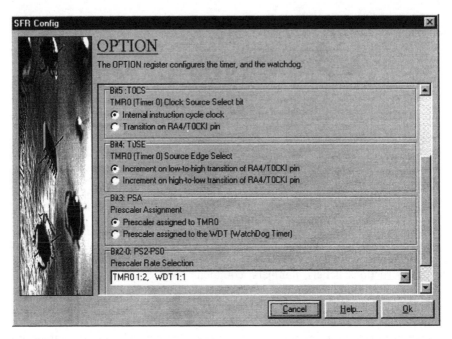

Figure 6.1 *A sample configuration window.*

For the beginner, there is an extensive help system with pages dedicated to all the PIC instructions, and all the different aspects of using PIC microcontrollers (e.g. subroutines, interrupts, EEPROM, etc.). The help system also helps the user with what to do with the program once it is finished.

For more up-to-date information on PIC Press, please consult this book's website: www.to-pic.com.

7
Sample programs

Program A

```
;**********************************
; written by: John Morton          *
; date: 21/07/97                   *
; version: 1.0                     *
; file saved as: LedOn             *
; for PIC16F54                     *
; clock frequency: 3.82 MHz        *
;**********************************

; PROGRAM FUNCTION: To turn on an LED.

        list        P=16F54
        include     "c:\pic\p16f5x.inc"

        __config    _RC_OSC & _WDT_OFF & _CP_OFF

;============
; Declarations:

porta   equ         05

        org         1FF
        goto        Start
        org         0

;============
; Subroutines:

Init    clrf        porta       ; resets Port A
        movlw       b'0000'     ; RA0: LED, RA1-3: not connected
        tris        porta
        retlw       0

;============
; Program Start:

Start   call        Init        ; sets up inputs and outputs
Main    bsf         porta, 0    ; turns on LED
```

```
                goto        Main           ; loops back to Main

                END
```

Program B

```
;***********************************
; written by: John Morton          *
; date: 21/07/97                   *
; version: 1.0                     *
; file saved as: PushButtonA       *
; for PIC16F54                     *
; clock frequency: 3.82 MHz        *
;***********************************
```

; PROGRAM FUNCTION: If a push button is pressed an LED is turned on.

```
                list        P=16F54
                include     "c:\pic\p16f5x.inc"

                __config    _RC_OSC & _WDT_OFF & _CP_OFF
```

```
;============
; Declarations:

porta    equ        05
portb    equ        06

         org        1FF
         goto       Start
         org        0
```

```
;============
; Subroutines:

Init     clrf       porta          ; resets inputs and outputs
         clrf       portb          ;
         movlw      b'0000'        ; RA0: LED, RA1-3: not connected
         tris       porta
         movlw      b'00000001'    ; RB0: push button, RB1-7: N/C
         tris       portb
         retlw      0
```

```
;============
; Program Start:

Start    call       Init

Main     btfss      portb, 0       ; tests push button, skip if pressed
         goto       LEDOff         ; push button isn't pressed so turns
```

```
                                 ;  LED off
              bsf      porta, 0   ; push button is pressed so turns
                                 ;  LED on
              goto     Main       ; loops back to Main
LEDOff        bcf      porta, 0   ; turns LED off
              goto     Main       ; loops back to Main

              END
```

Program C

```
;***********************************
; written by: John Morton          *
; date: 21/07/97                   *
; version: 2.0                     *
; file saved as: PushButtonB       *
; for PIC16F54                     *
; clock frequency: 3.82 MHz        *
;***********************************


; PROGRAM FUNCTION: If a push button is pressed an LED is turned on.

              list     P=16F54
              include  "c:\pic\p16f5x.inc"

              __config  _RC_OSC & _WDT_OFF & _CP_OFF

;============
; Declarations:

porta         equ      05
portb         equ      06

              org      1FF
              goto     Start
              org      0

;============
; Subroutines:

Init          clrf     porta      ; resets inputs and outputs
              clrf     portb      ;
              movlw    b'0000'    ; RA0: LED, RA1-3: not connected
              tris     porta
              movlw    b'00000001' ; RB0: push button, RB1-7: N/C
              tris     portb
              retlw    0
```

```
;============
; Program Start:

Start       call      Init

Main        movfw     portb    ; copies the number from Port B into
            movwf     porta    ;   the working register and then back
                               ;   into Port A

            goto      Main     ; loops back to Main

            END
```

Program D

```
;**********************************
; written by: John Morton          *
; date: 26/07/97                   *
; version: 1.0                     *
; file saved as: Timing            *
; for PIC16F54                     *
; clock frequency: 2.4576 MHz      *
;**********************************
```

; PROGRAM FUNCTION: The state of an LED is toggled every second and a
;   buzzer sounds for one second every five seconds.

```
            list      P=16F54
            include   "c:\pic\p16c5x.inc"

            __config  _XT_OSC & _WDT_OFF & _CP_OFF
```

```
;============
; Declarations:

porta       equ       05
portb       equ       06

Mark30      equ       08
Post80      equ       09
_5Second    equ       0A

            org       1FF
            goto      Start
            org       0
```

```
;============
; Subroutines:

Init        clrf      porta    ; resets inputs and outputs
            clrf      portb    ;
```

```
            movlw    b'0000'          ; RA0: LED, RA1-3: not connected
            tris     porta
            movlw    b'00000000'      ; RB0: buzzer, RB1-7: not connected
            tris     portb

            movlw    b'00000111'      ; sets up timing register
            option

            movlw    d'30'            ; sets up marker
            movwf    Mark30           ;

            movlw    d'80'            ; sets up first postscaler
            movwf    Post80           ;

            movlw    d'5'             ; sets up 5 seconds counter
            movwf    _5Second         ;

            retlw    0
```

```
;============
; Program Start:


Start       call     Init


Main        movfw    Mark30           ; takes the number out of Mark30
            subwf    TMR0, w          ; subtracts this number from the
                                      ;   number in TMR0, leaving the
                                      ;   result in the working register
                                      ;   (and leaving TMR0 unchanged)

            btfss    STATUS, Z        ; tests the Zero Flag - skip if set,
                                      ;   i.e. if the result is zero it will
                                      ;   skip the next instruction
            goto     Main             ; if the result isn't zero, it loops back to
                                      ;   'Loop'

            movlw    d'30'            ; moves the decimal number 30 into
            addwf    Mark30, f        ;   the w. reg. and then adds it to
                                      ;   'Mark30'

            decfsz   Post80, f        ; decrements 'Post80', and skips the
                                      ;   next instruction if the result is zero
            goto     Main             ; if the result isn't zero, it loops back to
                                      ;   'Loop'

                                      ; one second has now passed

            movlw    d'80'            ; resets postscaler
            movwf    Post80           ;
```

```
            comf      porta, f      ; toggles LED state
            bcf       portb, 0      ; turns off buzzer

            decfsz    _5Second, f   ; has five seconds passed?
            goto      Main          ; no, loop back

            bsf       portb, 0      ; turns on buzzer

            movlw     d'5'          ; resets 5 seconds counter
            movwf     _5Second      ;

            goto      Main          ; loops back to start

            END
```

Program E

```
;***********************************
; written by: John Morton            *
; date: 26/07/97                     *
; version: 1.0                       *
; file saved as: Traffic             *
; for PIC16F54                       *
; clock frequency: 2.4576 MHz        *
;***********************************
```

; PROGRAM FUNCTION: A pedestrian traffic lights junction is simulated.

```
            list      P=16F54
            include   "c:\pic\p16f5x.inc"

            __config  _XT_OSC & _WDT_OFF & _CP_OFF


;============
; Declarations:

porta       equ       05
portb       equ       06
Mark240     equ       08
PostX       equ       09
Counter8    equ       0A

            org       1FFh
            goto      Start
            org       0
```

```
;============
; Subroutines:

Init          clrf    porta           ; resets inputs and outputs
              clrf    portb           ;
              movlw   b'0001'         ; RA0: push button, RA1-3: N/C
              tris    porta           ;
              movlw   0               ; RB0-2: Motor. red, amber, green
              tris    portb           ; RB4, 5: Pedes. red, green

              movlw   b'00000111'     ; sets up timing register
              option                  ;

              retlw   0               ;

TimeDelay     movwf   PostX           ; sets up variable postscaler
              movlw   d'240'          ; sets up fixed marker
              movwf   Mark240         ;

TimeLoop      movfw   Mark240         ; waits for TMR0 to count up
              subwf   TMR0, w         ;   240 times
              btfss   STATUS, Z       ;
              goto    TimeLoop        ; hasn't, so keeps looping

              movlw   d'240'          ; resets Mark240
              addwf   Mark240, f      ;

              decfsz  PostX, f        ; does this X times
              goto    TimeLoop        ;

              retlw                   ; returns after required time


;============
; Program Start:

Start         call    Init            ;
Main          movlw   b'00010100'     ; motorists: green on, others off
              movwf   portb           ; pedestrians: red on, others off

ButtonLoop    btfss   porta, 0        ; is the pedestrians' button pressed?
              goto    ButtonLoop      ; no, so loops back

              bsf     portb, 1        ; motorists: amber on ...
              bcf     portb, 2        ; ... and green off

              movlw   d'20'           ; sends message of 2 seconds to sub
              call    TimeDelay       ; creates delay of required time

              movlw   b'00100001'     ; motorists: red on, amber off
              movwf   portb           ; pedestrians: green on, red off
```

```
            movlw      d'80'           ; sends message of 8 seconds to sub
            call       TimeDelay       ; creates delay of required time

            bsf        portb, 1        ; motorists: amber on . . .
            bcf        portb, 0        ; . . . and red off

            movlw      d'8'            ; sets up Counter8 with an initial
            movwf      Counter8        ;   value of 8

FlashLoop   movlw      d'5'            ; sends message of 0.5 second to sub
            call       TimeDelay       ; creates delay of required time
            movlw      b'00100010'     ; toggles the states of the lights
            xorwf      portb, f        ;
            decfsz     Counter8, f     ; runs through this loop 8 times
            goto       FlashLoop       ;
            goto       Main            ; loops back to start

            END
```

Program F

```
;***********************************
; written by: John Morton          *
; date: 17/08/97                   *
; version: 1.0                     *
; file saved as: Counter           *
; for PIC16F54                     *
; clock frequency: 3.82 MHz        *
;***********************************
```

; PROGRAM FUNCTION: To count the number of times a push button is
;   pressed, resetting after the sixteenth signal.

```
            list       P=16F54
            include    "c:\pic\p16f5x.inc"

            __config   _RC_OSC & _WDT_OFF & _CP_OFF

;============
; Declarations :

porta       equ        05
portb       equ        06
Counter     equ        08
```

```
            org     1FF
            goto    Start
            org     0
```

;============
; Subroutines:

```
Init        clrf    porta           ; resets I/O ports
            movlw   b'11111100'     ; moves the code for a 0 into Port B
            movwf   portb           ;

            movlw   b'0001'         ; RA0-3: not connected
            tris    porta

            movlw   b'00000000'     ; RB0: push button, RB1-7: 7-seg
            tris    portb           ;   code

            clrf    Counter         ; resets counter
            retlw   0

_7SegDisp   addwf   PCL             ; skips a certain number of
                                    ;   instructions
            retlw   b'11111110'     ; code for 0
            retlw   b'01100000'     ; code for 1
            retlw   b'11011010'     ; code for 2
            retlw   b'11110010'     ; code for 3
            retlw   b'01100110'     ; code for 4
            retlw   b'10110110'     ; code for 5
            retlw   b'10111110'     ; code for 6
            retlw   b'11100000'     ; code for 7
            retlw   b'11111110'     ; code for 8
            retlw   b'11110110'     ; code for 9
            retlw   b'11101110'     ; code for A
            retlw   b'00111110'     ; code for b
            retlw   b'10011100'     ; code for C
            retlw   b'01111010'     ; code for d
            retlw   b'10011110'     ; code for E
            retlw   b'10001110'     ; code for F
```

;============
; Program Start:

```
Start       call    Init            ; sets up inputs and outputs

Main        btfss   portb, 0        ; tests push button
            goto    Main            ; if not pressed, loops back

            incf    Counter         ;
            btfsc   Counter, 4      ; has Counter reached 16?
            clrf    Counter         ; if yes, resets Counter
```

```
                movfw    Counter      ; moves Counter into the working reg.
                call     _7SegDisp    ; converts into 7-seg code
                movwf    portb        ; displays value
                goto     Main         ; loops back to Main

                END
```

Program G

```
;**********************************
; written by: John Morton          *
; date: 17/08/97                   *
; version: 2.0                     *
; file saved as: Counter           *
; for PIC16F54                     *
; clock frequency: 3.82 MHz        *
;**********************************
```

; PROGRAM FUNCTION: To count the number of times a push button is
;   pressed, resetting after the sixteenth signal.

```
                list      P=16F54
                include   "c:\pic\p16f5x.inc"

                __config  _RC_OSC & _WDT_OFF & _CP_OFF


;============
; Declarations:

porta    equ     05
portb    equ     06
Counter  equ     08

                org     1FF
                goto    Start
                org     0


;============
; Subroutines:

Init            clrf     porta       ; resets I/O ports
                movlw    b'11111100'  ; moves the code for a 0 into Port B
                movwf    portb        ;
```

```
              movlw    b'0001'        ; RA0-3: not connected
              tris     porta
              movlw    b'00000000'    ; RB0: push button, RB1-7: 7-seg
              tris     portb          ; code

              clrf     Counter        ; resets counter
              retlw    0
_7SegDisp     addwf    PCL            ; skips a certain number of
                                      ;   instructions
              retlw    b'11111110'    ; code for 0
              retlw    b'01100000'    ; code for 1
              retlw    b'11011010'    ; code for 2
              retlw    b'11110010'    ; code for 3
              retlw    b'01100110'    ; code for 4
              retlw    b'10110110'    ; code for 5
              retlw    b'10111110     ; code for 6
              retlw    b'11100000'    ; code for 7
              retlw    b'11111110'    ; code for 8
              retlw    b'11110110'    ; code for 9
              retlw    b'11101110'    ; code for A
              retlw    b'00111110'    ; code for b
              retlw    b'10011100'    ; code for C
              retlw    b'01111010'    ; code for d
              retlw    b'10011110'    ; code for E
              retlw    b'10001110'    ; code for F


;============
; Program Start:

Start         call     Init           ; sets up inputs and outputs

Main          btfss    portb, 0       ; tests push button
              goto     Main           ; if not pressed, loops back

              incf     Counter        ;
              btfsc    Counter, 4     ; has Counter reached 16?
              clrf     Counter        ; if yes, resets Counter

              movfw    Counter        ; moves Counter into the working
                                      ;   reg.
              call     _7SegDisp      ; converts into 7-seg code
              movwf    portb          ; displays value
TestLoop      btfss    portb, 0       ; tests push button
              goto     Main           ; released, so loops back to Main
              goto     TestLoop       ; still pressed, so keeps looping

              END
```

Program H

```
;************************************
; written by: John Morton            *
; date: 17/08/97                     *
; version: 3.0                       *
; file saved as: Counter             *
; for PIC16F54                       *
; clock frequency: 3.82 MHz          *
;************************************


; PROGRAM FUNCTION: To count the number of times a push button is
;   pressed, resetting after the sixteenth signal.

            list        P=16F54
            include     "c:\pic\p16f5x.inc"

            __config    _RC_OSC & _WDT_OFF & _CP_OFF


;============
; Declarations:

porta     equ        05
portb     equ        06
Counter   equ        08

          org        1FF
          goto       Start
          org        0


;============
; Subroutines:

Init      clrf       porta            ; resets I/O ports
          movlw      b'11111100'      ; moves the code for a 0 into Port B
          movwf      portb            ;

          movlw      b'0001'          ; RA0-3: not connected
          tris       porta
          movlw      b'00000000'      ; RB0: push button, RB1-7: 7-seg
          tris       portb            ;   code

          movlw      b'00000111'      ; TMR0 prescaled by 256
          option                      ;
          clrf       Counter          ; resets counter

          retlw      0
```

```
_7SegDisp   addwf   PCL            ; skips a certain number of
                                   ;  instructions
            retlw   b'11111110'    ; code for 0
            retlw   b'01100000'    ; code for 1
            retlw   b'11011010'    ; code for 2
            retlw   b'11110010'    ; code for 3
            retlw   b'01100110'    ; code for 4
            retlw   b'10110110'    ; code for 5
            retlw   b'10111110     ; code for 6
            retlw   b'11100000'    ; code for 7
            retlw   b'11111110'    ; code for 8
            retlw   b'11110110'    ; code for 9
            retlw   b'11101110'    ; code for A
            retlw   b'00111110'    ; code for b
            retlw   b'10011100'    ; code for C
            retlw   b'01111010'    ; code for d
            retlw   b'10011110'    ; code for E
            retlw   b'10001110'    ; code for F

;============
; Program Start:

Start       call    Init           ; sets up inputs and outputs

Main        btfss   portb, 0       ; tests push button
            goto    Main           ; if not pressed, loops back

            incf    Counter        ;
            btfsc   Counter, 4     ; has Counter reached 16?
            clrf    Counter        ; if yes, resets Counter

            movfw   Counter        ; moves Counter into the working
                                   ;  reg.
            call    _7SegDisp      ; converts into 7-seg code
            movwf   portb          ; displays value

TestLoop    btfsc   portb, 0       ; tests push button
            goto    TestLoop       ; still pressed, so keeps looping

            clrf    TMR0           ; resets TMR0
TimeLoop    movlw   d'255'         ; has TMR0 reached 255?
            subwf   TMR0, w        ;
            btfss   STATUS, Z      ;
            goto    TimeLoop       ; if not, keeps looping
            goto    Main           ; 0.07 second has passed, so goes to
                                   ;  Main

            END
```

Program I

```
;**********************************
; written by: John Morton              *
; date: 10/01/05                       *
; version 1.0                          *
; file saved as StopClock              *
; for PIC16F54                         *
; clock frequency 2.4576 MHz           *
;**********************************
```

; PROGRAM FUNCTION: A stop clock displaying the time in tenths of
;   second, seconds, tens of seconds and minutes.

```
            list        P=16F54
            include     "c:\pic\p16f5x.inc"

            __config    _XT_OSC & _WDT_OFF & _CP_OFF
```

```
;============
; Declarations:
```

```
porta       equ     05
portb       equ     06
General     equ     08
Mark240     equ     09
Mark250     equ     0A
TenthSec    equ     0B
Seconds     equ     0C
TenSecond   equ     0D
Minutes     equ     0E

            #define     bounce      General, 1
            #define     start       General, 0

            org         1FFh
            goto        Start
            org         0
```

```
;============
; Subroutines:
```

```
Init        clrf        porta           ; resets input/output ports
            clrf        portb           ;
            movlw       b'0000'         ; bits 0-3: 7-seg display
            tris        porta           ;   control
            movlw       b'00000001'     ; bits 1-7: 7-seg display code
            tris        portb           ; bit 0: start/stop button
```

```
            movlw    b'00000111'   ; TMR0 prescaled by 256
            option

            clrf     TenthSec      ; resets timing registers
            clrf     Seconds       ;
            clrf     TenSecond     ;
            clrf     Minutes       ;

            bcf      start         ; initially, stop state
            bsf      bounce        ; bounce initially set

            movlw    d'240'        ; sets up marker register
            movwf    Mark240       ;
            retlw    0             ;
```

;================================
Debounce

```
            movfw    Mark250       ; if about 0.1 second has
            subwf    TMR0, w       ;   passed, sets the bounce
            btfss    STATUS, Z     ;   bit
            retlw    0             ;
            bsf      bounce        ;
            retlw    0
```

PrimeBounce

```
            bcf      bounce        ; clears bounce bit to trigger
            movlw    d'250'        ;   and sets up Mark250 so that
            addwf    TMR0, w       ;   about 0.1 second will be
            movwf    Mark250       ;   counted
            retlw    0             ;
```

;================================
```
Update      btfsc    start         ; checks start/stop state
            call     Timing        ; if start, updates timers

            btfss    bounce        ; checks whether or not to test
            call     Debounce      ;   whether 0.1 second has passed?

            movlw    b'00000011'   ; ignores all but bits 0 and 1
            andwf    TMR0, w       ;   of TMR0, leaving result in w
            addwf    PCL, f        ; adds result to PC, in order to
            goto     Display10th   ;   select a display
            goto     Display1      ;
            goto     Display10     ;
            goto     DisplayMin    ;
```

Display10th
```
            movfw    TenthSec      ; takes the number from TenthSec
            call     _7SegDisp     ; converts number into 7-seg code
```

```
                movwf    portb         ; displays value through Port B
                movlw    b'0010'       ; turns on correct display
                movwf    porta         ;
                retlw    0             ; returns
```

Display1

```
                movfw    Seconds       ; takes the number from Seconds
                call     _7SegDisp     ; converts number into 7-seg code
                movwf    portb         ; displays value through Port B
                movlw    b'0001'       ; turns on correct display
                movwf    porta         ;
                retlw    0             ; returns
```

Display10

```
                movfw    TenSecond     ; takes the number from TenSeconds
                call     _7SegDisp     ; converts number into 7-seg code
                movwf    portb         ; displays value through Port B
                movlw    b'1000'       ; turns on correct display
                movwf    porta         ;
                retlw    0             ; returns
```

DisplayMin

```
                movfw    Minutes       ; takes the number from Minutes
                call     _7SegDisp     ; converts number into 7-seg code
                movwf    portb         ; displays value through Port B
                movlw    b'0100'       ; turns on correct display
                movwf    porta         ;
                retlw    0             ; returns
```

```
;================================
```

_7SegDisp

```
                addwf    PCL, f        ; returns with correct 7-seg code
                retlw    b'11111100'   ; code for 0
                retlw    b'01100000'   ; code for 1
                retlw    b'11011010'   ; code for 2
                retlw    b'11110010'   ; code for 3
                retlw    b'01100110'   ; code for 4
                retlw    b'10110110'   ; code for 5
                retlw    b'10111110'   ; code for 6
                retlw    b'11100000'   ; code for 7
                retlw    b'11111110'   ; code for 8
                retlw    b'11110110'   ; code for 9
```

```
;================================
```

```
Timing          movfw    Mark240       ; tests to see if 0.1 second has
                subwf    TMR0, w       ;  passed
```

```
        btfss    STATUS, Z      ;
        retlw    0              ; 0.1 second hasn't passed - returns

        movlw    d'240'         ; updates Mark240
        addwf    Mark240,f      ;

        incf     TenthSec, f    ; adds 1 to number of 0.1 second

        movlw    d'10'          ; tests to see whether TenthSec has
        subwf    TenthSec, w    ;   reached 10 (has one second passed?)
        btfss    STATUS, Z      ;
        retlw    0              ; 1 second hasn't passed, so returns

        clrf     TenthSec       ; 1 second has passed, so resets
        incf     Seconds, f     ;   TenthSec and adds 1 to Seconds

        movlw    d'10'          ; tests to see whether Seconds has
        subwf    Seconds, w     ;   reached 10 (whether ten seconds
        btfss    STATUS, Z      ;   has passed)
        retlw    0              ;

        clrf     Seconds        ; 10 seconds have passed, so resets
        incf     TenSecond,f    ;   Seconds and adds 1 to TenSecond

        movlw    d'6'           ; tests to see whether TenSecond has
        subwf    TenSecond, w   ;   reached 6 (whether one minute
        btfss    STATUS, Z      ;   has passed)
        retlw    0              ;

        clrf     TenSecond      ; 60 seconds have passed, so resets
        incf     Minutes, f     ;   TenSecond and adds 1 to Minutes

        movlw    d'10'          ; tests to see whether Minutes has
        subwf    Minutes, w     ;   reached 10 (whether ten minutes
        btfss    STATUS, Z      ;   has passed)
        retlw    0              ;

        clrf     Minutes        ; 10 minutes have passed, so resets
        retlw    0              ;
```

```
;=====================================
Start     call    Init           ; runs initialisation routine

Released  call    Update         ; keeps timing and display updated
          btfss   bounce         ; waits 0.1 s to confirm button is
          goto    Released       ;   released
          btfss   portb, 0       ; has button now been pressed?
          goto    Released       ; no, so keeps looping

          movlw   b'00000001'    ; toggles state of start/stop bit
          xorwf   General, f     ;
          call    PrimeBounce    ; primes de-bounce routine
```

```
Pressed   call      Update        ; keeps timing and display updated
          btfss     bounce        ; waits 0.1 s to confirm button is
          goto      Pressed       ;  pressed
          btfsc     portb, 0      ; has button now been released?
          goto      Pressed       ; no, so keeps looping

          call      PrimeBounce   ; primes de-bounce routine
          goto      Released      ;

          END
```

Program J

```
;**********************************
; written by: John Morton          *
; date: 21/07/97                   *
; version: 1.0                     *
; file saved as: LogicGates        *
; for PIC16F54                     *
; clock frequency: 3.82 MHz        *
;**********************************
```

; PROGRAM FUNCTION: To act as the eight different gates.

```
          list      P=16F54
          include   "c:\pic\p16f5x.inc"


          __config  _RC_OSC & _WDT_OFF & _CP_OFF


;============
; Declarations:

porta     equ       05
portb     equ       06
STORE     equ       08

          org       1FF
          goto      Start
          org       0


;============
; Subroutines:

Init      movlw     b'1111'       ; RA0: secondary input, RA1-3: gate
```

```
          tris      porta        ; select bits
          movlw     b'00000001'  ; RB0: primary input, RB4:
                                 ;  output,
          tris      portb        ; RB1-3 and RB5-7: not
                                 ;  connected
          retlw     0
```

```
;============
; Program Start:
```

```
Start       call      Init         ; sets up inputs and outputs

Main        bcf       STATUS, C    ; makes sure carry flag is clear
            rrf       porta, w     ; bumps off bit 0, leaving result
                                   ;  in w
            andlw     b'0011'      ; masks all but bits 0 and 1
            addwf     PCL, f       ; branches accordingly
            goto      BufferNOT    ; handles Buffers and NOTs
            goto      ANDNAND      ; handles ANDs and NANDs
            goto      IORNOR       ; handles IORs and NORs

XORXNOR     movfw     porta        ; takes Input B
            xorwf     portb,w      ; and XORs with Input A
Common      movwf     STORE        ; stores result
            btfsc     porta, 3     ; tests inversion bit
            comf      STORE, f     ; inverts output if necessary
            swapf     STORE, w     ; moves result into bit 4
            movwf     portb        ; outputs result
            goto      Main         ;

BUFFERNOT
            movfw     portb        ; takes Input A unchanged
            goto      Common       ; rest is as in XOR/XNOR section

ANDNAND
            movfw     porta        ; takes Input B
            andwf     portb,w      ;   and ANDs with Input A
            goto      Common       ; rest is as in XOR/XNOR section

IORNOR      movfw     porta        ; takes Input B
            iorwf     portb,w      ;   and IORs with Input A
            goto      Common       ; rest is as in XOR/XNOR section

            END
```

Program K

```
;**********************************
; written by: John Morton          *
; date: 07/08/97                   *
; version: 1.0                     *
; file saved as: Alarm             *
; for PIC16F54                     *
; clock frequency: 3.82 MHz        *
;**********************************
```

; PROGRAM FUNCTION: An alarm system which can be set or disabled.

```
                list        P=16F54
                include     "c:\pic\p16f5x.inc"


                __config    _RC_OSC & _WDT_ON & _CP_OFF
```

```
;============
; Declarations:
```

```
porta           equ         05
portb           equ         06
Post50          equ         08
Counter         equ         09

                org         1FF
                goto        Start
                org         0
```

```
;============
; Subroutines:
```

```
Init            clrf        porta       ; resets inputs and outputs
                clrf        portb       ;
                movlw       b'0011'     ; RA0: Sensor, RA1: Settings
                                        ;   switch
                tris        porta       ; RA2: not connected, RA3:
                                        ;   siren
                movlw       b'00000000' ; RB0: green LED, RB1: red LED
                tris        portb       ;   RB2-7: not connected

                movlw       b'00001111' ; WDT prescaled by 128 (TMR0
                option                  ;   not prescaled)
```

```
                  clrf     Counter        ; resets clock cycle counter
                  movlw    d'50'          ; sets up postscaler
                  retlw    0
```

```
;============
; Program Start:

Start             call     Init

Main              btfsc    porta, 1       ; tests setting switch
                  goto     GreenLed       ; switch is high, so turn on green
                                          ;   LED
                  bsf      portb, 1       ; switch is low, so turn on red LED

TenthSecond
                  decfsz   Counter        ; has 1/10th second passed?
                  goto     Continue       ;
                  decfsz   Post50         ;
                  goto     Continue       ;
                  clrf     portb          ; it has, so turns off all LEDs

Continue          btfsc    porta, 1       ; tests setting switch
                  goto     Waste2Cycle    ; disabled, so doesn't test trigger
                                          ;   input
                  btfss    porta, 0       ; has motion sensor been set?
                  goto     TenthSecond    ; not triggered, so loops back

                  bsf      porta, 3       ; turns on siren
EndLoop           clrwdt                  ; resets watchdog timer
                  goto     EndLoop        ; constantly loops

GreenLed          bsf      portb, 0       ; turns on green LED
                  goto     TenthSecond    ; loops back to main body of
                                          ;   program

Waste2Cycle       goto     TenthSecond    ; wastes two clock cycles

                  END
```

Program L

```
;***********************************
; written by: John Morton          *
; date: 24/08/97                   *
; version: 1.0                     *
; file saved as: Bike              *
; for PIC16F54                     *
; clock frequency: 2.4576 MHz      *
;***********************************
```

; PROGRAM FUNCTION: A bicycle speedometer and mileometer.

```
            list        P=16F54
            include     "c:\pic\p16f5x.inc"

            __config    _RC_OSC & _WDT_OFF & _CP_OFF
```

;============
; Declarations:

```
porta       equ         05
portb       equ         06

Dist1       equ         09
Dist10      equ         0B
Dist100     equ         08

SP10th      equ         0D
SP1         equ         0F
SP10        equ         0C

Speed10th   equ         10
Speed1      equ         11
Speed10     equ         12

General     equ         13
Mark89      equ         14
tempa       equ         15
_10         equ         16

            #define     mode        portb, 0
            #define     counter     porta, 3
            #define     debouncer   General, 0

            org         1FFh
            goto        Start
            org         0
```

;============
; Subroutines:

```
Init        movlw       b'0001'     ; yes, so resets Port A
            movwf       porta       ;
            clrf        portb       ;
            movlw       b'1000'     ; RA0-2: controllers for 7-seg
            tris        porta       ;   display, RA3 - counter
```

```
                movlw   b'00000001'   ; RB0: select switch, RB1-7 7-
                tris    portb         ;   seg code

                movlw   d'9'          ; resets speed regs.
                movwf   Speed10th     ;
                movwf   Speed1        ;
                movwf   Speed10       ;

                clrf    Dist1
                clrf    Dist10
                clrf    Dist100
                clrf    TMR0          ;
                clrf    SP1
                clrf    SP10th
                clrf    SP10
                retlw   0

Display         movwf   FSR           ; speed, or distance
                decfsz  _10           ; changes display every ten times
                retlw   0             ;   it gets here
                movlw   d'10'         ;
                movwf   _10           ;

                movlw   b'0111'
                andwf   porta, w
                movwf   tempa
                bcf     STATUS, C
                rrf     tempa         ; selects next display
                btfss   STATUS, C
                goto    CodeSelect
                movlw   b'0100'       ; yes, so resets Port A
                movwf   tempa         ;

CodeSelect      movlw   b'0111'       ; ignores button
                andwf   porta,w       ; uses Port A to select correct
                addwf   FSR, f        ;   file register
                movfw   INDF          ; takes out the correct code
                call    _7SegDisp     ; converts code
                movwf   portb         ; displays number
                movfw   tempa
                movwf   porta
                retlw   0             ; returns

_7SegDisp       addwf   PCL           ; returns with correct code
                retlw   b'01111110'   ; 0
```

```
                retlw    b'00001100'    ; 1
                retlw    b'10110110'    ; 2
                retlw    b'10011110'    ; 3
                retlw    b'11001100'    ; 4
                retlw    b'11011010'    ; 5
                retlw    b'11111010'    ; 6
                retlw    b'00001110'    ; 7
                retlw    b'11111110'    ; 8
                retlw    b'11011110'    ; 9
                retlw    b'01110000'    ; L

Debounce        btfsc    debouncer      ; has signal finished?
                goto     NextTest       ; yes, so tests button

                btfss    counter        ; has signal finished?
                bsf      debouncer      ; yes, so sets bit
                retlw    0              ; no, so returns

NextTest        btfss    counter        ; second signal?
                retlw    0              ; no, so returns

                movfw    Speed10th      ; transfers file regs. so that
                movwf    SP10th         ;   values are displayed
                movfw    Speed1         ;
                movwf    SP1            ;
                movfw    Speed10        ;
                movwf    SP10           ;

                movlw    d'9'           ; resets speed regs.
                movwf    Speed10th      ;
                movwf    Speed1         ;
                movwf    Speed10        ;
                bcf      debouncer
                retlw    0

;=============
; Program Start:

Start           call     Init

Main            btfsc    mode           ; which mode is it in?
                goto     Speed          ; Speed mode

;=========
Distance        movlw    b'00110100'    ; TMR0 counts external signals
                option                  ; prescaled by 32
```

```
DistLoop        btfsc    mode              ; checks mode
                goto     Speed             ; Speed mode

                movlw    07h
                call     Display           ;

                movlw    d'21'             ; has TMR0 reached 21?
                subwf    TMR0, w           ;

                btfss    STATUS, Z         ;
                goto     DistLoop          ; no, so loops back

                incf     Dist1             ; increments 1 kms
                clrf     TMR0

                movlw    d'10'             ; has Dist1 reached 10?
                subwf    Dist1, w          ;
                btfss    STATUS, Z         ;
                goto     DistLoop          ; no, so loops back

                incf     Dist10            ; increments 10 kms
                clrf     Dist1

                movlw    d'10'             ; has Dist10 reached 10?
                subwf    Dist10, w         ;
                btfss    STATUS, Z         ;
                goto     DistLoop          ; no, so loops back

                incf     Dist100           ; increments 100 kms
                clrf     Dist10

                movlw    d'10'             ; has Dist100 reached 10?
                subwf    Dist100, w        ;
                btfss    STATUS, Z         ;
                goto     DistLoop          ; no, so loops back
                clrf     Dist100           ; has passed limit, so resets and
                goto     Main              ;   loops back

;========
Speed           movlw    b'00000110'       ; TMR0: internal, prescaled
                option                     ;   at 128

                btfss    counter           ; waits for first signal
                goto     Speed+2           ; keeps looping

BasicTimeLoop
                btfss    mode              ; checks mode
                goto     Distance          ; Speed mode
```

```
                movlw   0Bh             ;
                call    Display         ;

                call    Debounce        ;

                movfw   Mark89          ; has 0.0185 second passed?
                subwf   TMR0, w         ;
                btfss   STATUS, Z       ;
                goto    BasicTimeLoop   ; no, so loops back

                movlw   d'89'           ; (adds 89 to marker)
                addwf   Mark89          ;

                decf    Speed10th, f    ; yes, so decrements speed by
                                        ;   one tenth of a km per hour
                movlw   d'255'          ; has it passed 0?
                subwf   Speed10th, w    ;
                btfss   STATUS, Z       ;
                goto    BasicTimeLoop   ; no, so loops back

                movlw   d'9'            ; resets 10th unit
                movwf   Speed10th       ;
                decf    Speed1, f       ;
                movlw   d'255'          ; has it passed 0?
                subwf   Speed1, w       ;
                btfss   STATUS, Z       ;
                goto    BasicTimeLoop   ;

                movlw   d'9'            ; resets 1 unit
                movwf   Speed1          ;
                decf    Speed10, f      ;
                movlw   d'255'          ; has it passed 0?
                subwf   Speed10, w      ;
                btfss   STATUS, Z       ;
                goto    BasicTimeLoop   ;

TooSlow         clrf    SP10th          ; displays "SLO" on the displays
                movlw   d'10'
                movwf   SP1
                movlw   d'5'
                movwf   SP10
                movlw   0Bh             ;
                call    Display         ;
                btfss   counter         ; tests for button
                goto    TooSlow         ; no, so keeps looping

                movlw   d'9'            ; resets speed regs.
                movwf   Speed10         ;
```

```
        goto        BasicTimeLoop ;

        END
```

Program M

```
;*****************************
; written by: John Morton        *
; date: 10/01/05                 *
; version: 1.0                   *
; file saved as: dice.asm        *
; for P12F508                    *
; clock frequency: Int. 4 MHz    *
;*****************************

; PROGRAM FUNCTION: A pair of dice.

        list        P=12F508
        include     "c:\pic\p12f508.inc"

        __config    _MCLRE_OFF & _CP_OFF & _WDT_OFF &
                    _IntRC_OSC

;===============
; Declarations:

Die1num   equ       10h
Die2num   equ       11h
Mark60    equ       12h
PostX     equ       13h
PostVal   equ       14h
Ran1      equ       15h
Ran2      equ       16h
General   equ       17h
Random    equ       18h

#define   slow      General, 0

          org       0           ; first instruction to be executed
          movwf     OSCCAL      ; calibrates internal oscillator
          goto      Start       ;

;==========================================
; Subroutines:

Init      movlw     b'100000'   ; turns off all LEDs
          movwf     GPIO        ;
```

```
        movlw   b'001000'      ; sets up which pins are inputs
        tris    GPIO           ;   and which are outputs

        movlw   b'01000111'    ; enable wake-on-change, disable weak
        option                 ;   pull-ups, TMR0 prescaled by 256

        movlw   d'4'           ; sets up postscalers
        movwf   PostX          ;
        movwf   PostVal        ;

        clrf    Die1num        ; clears display registers
        clrf    Die2num        ;
        clrf    Ran1           ; clears random number registers
        clrf    Ran2           ;
        bcf     slow           ; clears 'slow-down' flag
        retlw   0              ;

;=====================================
Display btfss   TMR0, 4        ; uses bit 4 of TMR0 to choose die
        goto    Die2           ;

        movfw   Die1num        ; gets number to display
        call    Code1          ; converts to code
        movwf   GPIO           ; outputs
        retlw   0              ;

Die2    movfw   Die2num        ; gets number to display
        call    Code2          ; converts to code
        movwf   GPIO           ; outputs
        retlw   0              ;

; arrangement for dice 1 is : CTLR, D, -, A, C, B

Code1   addwf   PCL, f         ;
        retlw   b'100000'      ; all off
        retlw   b'100100'      ; 1
        retlw   b'100001'      ; 2
        retlw   b'100101'      ; 3
        retlw   b'100011'      ; 4
        retlw   b'100111'      ; 5
        retlw   b'110011'      ; 6
        retlw   b'110111'      ; all on

; arrangement for dice 2 is : CTLR, C, -, B, A, D

Code2   addwf   PCL, f         ;
        retlw   b'010111'      ; all off
        retlw   b'010101'      ; 1
        retlw   b'010011'      ; 2
```

```
            retlw    b'010001'      ; 3
            retlw    b'000011'      ; 4
            retlw    b'000001'      ; 5
            retlw    b'000010'      ; 6
            retlw    b'000000'      ; all on

;=====================================
Timing      movfw    Mark60         ; 1/40th second delay
            subwf    TMR0, w        ;
            btfss    STATUS, Z      ;
            retlw    0              ;

            movlw    d'60'          ; resets marker
            addwf    Mark60, f      ;

            decfsz   PostX, f       ; variable further delay
            retlw    0              ;

            call     RandomGen      ; generate new pseudo-random
            swapf    Random, w      ;   number
            andlw    b'00000111'    ; converts to 0-7 and moves
            movwf    Die1num        ;   into Die1num

            call     RandomGen      ; generate new pseudo-random
            swapf    Random, w      ;   number
            andlw    b'00000111'    ; converts to 0-7 and moves
            movwf    Die2num        ;   into Die2num

            btfsc    slow           ; should this slow down?
            call     Slowdown       ; yes

            movfw    PostVal        ; updates variable delay length
            movwf    PostX          ;
            retlw    0

Slowdown    incf     PostVal, f     ; increases delay length
            btfsc    PostVal, 5     ; has PostVal reached 32?
            clrf     PostVal        ; resets, telling 'Released' section
            retlw    0              ;   that the dice have stopped rolling

RandomGen
            movlw    d'63'          ; newRandom =
            addwf    Random, w      ;   63 + oldRandom x 3
            addwf    Random, w      ;
            addwf    Random, f      ;
            retlw    0              ;
```

```
;======================================
RandomScroll
        incf    Ran1, f         ; v. quickly scrolls through
        movlw   d'6'            ;   has Ran1 reached 6?
        subwf   Ran1, w         ;
        btfss   STATUS, Z       ;
        retlw   0               ; no, so returns

        clrf    Ran1            ;
        incf    Ran2, f         ;
        movlw   d'6'            ; has Ran1 reached 6?
        subwf   Ran2, w         ;
        btfss   STATUS, Z       ;
        retlw   0               ; no, so returns
        clrf    Ran2            ;
        retlw   0               ;

;======================================
; PROGRAM START

Start   call    Init            ; initial settings

Pressed btfsc   GPIO, 3         ; tests button
        goto    Released        ; branches when released
        call    RandomScroll    ; quickly scrolls through no.s
        call    Timing          ; keeps flashing going
        call    Display         ; keeps displays changing
        goto    Pressed         ;

Released bsf    slow            ; tells Timing to slow down
        call    Timing          ; keeps flashing going
        call    Display         ; keeps displays going
        movf    PostVal, f      ; have dice stopped rolling?
        btfss   STATUS, Z       ;
        goto    Released+1      ; no, so keeps looping

        incf    Ran1, w         ; moves 1+ the random number
        movwf   Die1num         ;   into the display regs.
        incf    Ran2, w         ;
        movwf   Die2num         ;

        movlw   d'240'          ; 240 x 1/40th second = 6 second
        movwf   PostX           ;   delay

EndLoop call    Display         ; 6 second delay, after which all
        movfw   Mark60          ;   LEDs are turned off
        subwf   TMR0, w         ;
```

```
        btfss       STATUS, Z        ;
        goto        EndLoop          ;

        movlw       d'60'            ;
        addwf       Mark60, f        ;

        decfsz      PostX, f         ;
        goto        EndLoop          ;

        movlw       b'100000'        ; turns off all LEDs
        movwf       GPIO             ;
        sleep                        ; goes to low power mode

        END
```

Program N

```
;****************************************
; written by: John Morton              *
; date: 14/03/05                       *
; version: 1.0                         *
; file saved as: quiz.asm              *
; for PIC12F675                        *
; clock frequency: Int. 4 MHz          *
;****************************************

; Program Description: Quiz controller for 3 players, including reset
;   button for the quiz master.

        list        P=12F675
        include     "c:\pic\p12f675.inc"

;==============
; Declarations:

temp    equ         20h
Post16  equ         21h

        org         0                ; first instruction to be executed
        goto        Start            ;

        org         4                ; interrupt service routine
        goto        isr              ;

;==============
; Subroutines:

Init    bsf         STATUS, RP0      ; goes to Bank 1
        call        3FFh             ; calls calibration address
```

```
        movwf    OSCCAL            ; moves w. reg into OSCCAL

        movlw    b'011110'         ; GP5: Buzzer, GP3: Reset button
        movwf    TRISIO            ; GP1,2,4: LEDs/Buttons (inputs
                                   ;   to start with), GP0: LED enable
        movlw    b'010110'         ; GP1,2,4 have weak pull-ups
        movwf    WPU               ;   enabled

        movlw    b'00000111'       ; pull-ups enabled, TMR0 presc.
        movwf    OPTION_REG        ;   by maximum amount (256)
        clrf     PIE1              ; turns off peripheral interrupts
        movlw    b'010110'         ; enables GPIO change interrupt
        movwf    IOC               ;   on GP1, GP2 and GP4 only
        clrf     VRCON             ; turns off comparator V. ref.
        clrf     ANSEL             ; makes GP0:3 digital I/O pins
        bcf      STATUS, RP0       ; back to Bank 0
        clrf     GPIO              ; resets input/output port
        movlw    b'00001000'       ; enables GPIO change interrupt
        movwf    INTCON            ;   only
        movlw    b'00000111'       ; turns off comparator
        movwf    CMCON             ;
        clrf     T1CON             ; turns off TMR1
        clrf     ADCON0            ; turns off A to D converter

        movlw    d'16'             ; sets up postscaler
        movwf    Post16            ;
        retfie                     ; returns, enabling interrupts

;=====================
; Interrupt Service Routine
isr     btfss    INTCON, 0         ; checks GPIO change int. flag
        goto     Timer             ; TMR0 interrupt occurred ...
                                   ; GPIO interrupt occurred ...
        bcf      INTCON, 0         ; resets interrupt flag

        comf     GPIO, w           ; stores state of GPIO
        andlw    b'010110'         ; masks all except buttons
        movwf    temp              ;
        btfsc    STATUS, Z         ; are any buttons actually pressed?
        retfie                     ; false alarm

        bsf      STATUS, RP0       ; moves to Bank 1
        movlw    b'001000'         ; makes GP1,2,4 outputs
        movwf    TRISIO            ;
        bcf      STATUS, RP0       ; moves to Bank 0
```

```
            movfw     temp              ; moves temp back into GPIO,
            addlw     b'100001'         ;   sets GP5 and GP0 (turns on
            movwf     GPIO              ;   buzzer and enables LEDs)

            movlw     b'00100000'       ; enables TMR0 interrupt, disables
            movwf     INTCON            ;   the GPIO change interrupt
            retfie                      ; returns, enabling GIE

Timer       bcf       INTCON, 2         ; resets TMR0 interrupt flag
            decfsz    Post16, f         ; is this the 16th TMR0 interrupt
            retfie                      ;

            bcf       GPIO, 5           ; turn off buzzer
            clrf      INTCON            ; turns off all interrupts
            sleep                       ; goes into low power mode

;=============
; Program Start

Start       call      Init              ; initialisation routine

Main        goto      Main              ; keeps looping

            END
```

Program O

```
;************************************ ***
; written by: John Morton                 *
; date: 10/01/05                          *
; version 1.0                             *
; file saved as phonecard.asm             *
; for P12F675                             *
; clock frequency: internal 4 MHz         *
; *************************************

; Program Description: A smart card for a phone box.

            list          P=12F675
            include       "c:\pic\p12f675.inc"


;====================
; Declarations:

W_temp        equ       20h
STATUS_temp   equ       21h
temp          equ       22h
```

```
Mark125    equ    23h
Post125    equ    24h
Post15     equ    25h

           org    0               ; first instruction to be executed
           goto   Start           ;
           org    4               ; interrupt service routine
           goto   isr             ;
```

```
;============
; Subroutines:
```

```
Init       bsf    STATUS, RP0     ; goes to Bank 1
           call   3FFh            ; calls calibration address
           movwf  OSCCAL          ; moves w. reg into OSCCAL
           movlw  b'111110'       ; all inputs except GP0
           movwf  TRISIO          ;
           clrf   WPU             ; weak pull-ups disabled
           movlw  b'11000111'     ; sets up timer and some pin
           movwf  OPTION_REG      ;   settings
           clrf   PIE1            ; turns off peripheral ints.
           clrf   IOC             ; disables GPIO change int.
           clrf   VRCON           ; turns off comparator V. ref.
           clrf   ANSEL           ; makes GP0:3 digital I/O pins

           bcf    STATUS, RP0     ; back to Bank 0
           clrf   GPIO            ; resets input/output port
           movlw  b'00010000'     ; sets up interrupts
           movwf  INTCON          ;
           movlw  b'00000111'     ; turns off comparator
           movwf  CMCON           ;
           clrf   T1CON           ; turns off TMR1
           clrf   ADCON0          ; turns off A to D conv.

           movlw  d'125'          ; sets up postscalers
           movwf  Post125         ;
           movlw  d'15'           ;
           movwf  Post15          ;
           retfie                 ; returns from Init

isr        movwf  W_temp          ; stores w. reg in temp register
           movfw  STATUS          ; stores STATUS in temp
           movwf  STATUS_temp     ;   register

           bcf    INTCON, 1       ; resets INT interrupt flag
           bcf    STATUS, RP0     ; makes sure we're in Bank 0
```

```
                movfw    GPIO              ; reads value of GPIO
                movwf    temp              ;
                rrf      temp, f           ; rotates right three times ...
                rrf      temp, f           ;
                rrf      temp, w           ; ... leaving result in w. reg
                andlw    b'000111'         ; masks bits 3-5
                call     CardValue         ; converts code into minutes

                bsf      STATUS, RP0       ; goes to Bank 1
                movwf    EEDATA            ; stores minutes in EEDATA
                clrf     EEADR             ; selects EEPROM address 00h
                bsf      EECON1, 2         ; enables a write operation
                movlw    0x55              ; now follows the 'safe
                movwf    EECON2            ;   combination'
                movlw    0xAA              ;
                movwf    EECON2            ;
                bsf      EECON1, 1         ; starts the write operation
EELoop          btfsc    EECON1, 1         ; has write operation finished?
                goto     EELoop            ; no, still high, so keeps looping

                movfw    STATUS_temp       ; restores STATUS register to
                movwf    STATUS            ;   original value
                swapf    W_temp, f         ; restores working register to
                swapf    W_temp, w         ;   original value
                retfie                     ; returns, enabling GIE

CardValue       addwf    PCL, f            ; returns with new number of
                retlw    d'2'              ;   minutes for the card
                retlw    d'5'              ;
                retlw    d'10'             ;
                retlw    d'20'             ;
                retlw    d'40'             ;
                retlw    d'60'             ; one hour
                retlw    d'120'            ; two hours
                retlw    0                 ; (erases card)


;===========
; Program Start

Start           call     Init              ; initialisation routine

Main            bsf      STATUS, RP0       ; selects Bank 1
                clrf     EEADR             ; selects EEPROM address 00h
                bsf      EECON1, 0         ; initiates an EEPROM read
                movfw    EEDATA            ; reads EEDATA
                bcf      STATUS, RP0       ; selects Bank 0
```

```
              btfss     STATUS, Z        ; is it 0?
              goto      Active           ; no, so goes to Active
              bcf       GPIO, 0          ; turns off GP0
              sleep                      ; goes to sleep
              nop                        ;
              goto      Main             ; loops back to Main

Active        bsf       GPIO, 0          ; turns on GP0
              btfss     GPIO, 1          ; is a call in progress?
              goto      Active           ; no, so keeps waiting
              movfw     Mark125          ; has one minute passed?
              subwf     TMR0, w          ;
              btfss     STATUS, Z        ;
              goto      Active           ; no, so keeps looping

              movlw     d'125'           ;
              addwf     Mark125          ;
              decfsz    Post125          ;
              goto      Active           ;

              movlw     d'125'           ;
              movwf     Post125          ;
              decfsz    Post15           ;
              goto      Active           ;

              movlw     d'15'            ; one minute has passed, so
              movwf     Post15           ;   resets final postscaler
              bsf       STATUS, RP0      ; goes to Bank 1
              clrf      EEADR            ; selects EEPROM address 00h
              bsf       EECON1, 0        ; reads EEPROM address 00h
              decf      EEDATA           ; subtracts 1 minute from card
              bsf       EECON1, 2        ; enables a write operation
              bcf       INTCON, 7        ; disables global interrupts
              movlw     0x55             ; now follows the 'safe
              movwf     EECON2           ;   combination'
              movlw     0xAA             ;
              movwf     EECON2           ;
              bsf       EECON1, 1        ; starts the write operation
EELoop        btfsc     EECON1, 1        ; has write operation finished?
              goto      EELoop           ; no, still high, so keeps looping

              bcf       STATUS, RP0      ; back to Bank 0
              bsf       INTCON, 7        ; enables global interrupts
              goto      Main             ; loops back to start

              END
```

Program P

```
;**************************************
; written by: John Morton              *
; date: 14/03/05                       *
; version: 1.0                         *
; file saved as: tempsense.asm         *
; for PIC12F675                        *
; clock frequency: Int. 4 MHz          *
; **************************************

; Program Description: Bath temperature measuring device.

        list      P=12F675
        include   "c:\pic\p12f675.inc"

;============
; Declarations:

W_temp     equ     20h
STATUS_temp
           equ     21h

           org     0               ; first instruction to be executed
           goto    Start           ;

           org     4               ; interrupt service routine
           goto    isr             ;

;============
; Subroutines:

Init       bsf     STATUS, RP0     ; goes to Bank 1
           call    3FFh            ; calls calibration address
           movwf   OSCCAL          ; moves w. reg into OSCCAL
           movlw   b'010000'       ; GP0-2 are LEDs, GP4 analogue
           movwf   TRISIO          ;   input
           clrf    WPU             ; weak pull-ups disabled

           movlw   b'10000000'     ; weak pull-ups disabled, no timer
           movwf   OPTION_REG      ;   used
           movlw   b'01000000'     ; enables A/D interrupt
           movwf   PIE1            ;
           clrf    IOC             ; disables GPIO change int.
           clrf    VRCON           ; turns off comparator V. ref.
           movlw   b'00011000'     ; A/D clock: Fosc/8 = 2 µs;
                                   ;   AN3/GP4
           movwf   ANSEL           ; is anal. input, others are digital
```

```
        bcf      STATUS, RP0      ; back to Bank 0
        clrf     GPIO             ; resets input/output port
        movlw    b'01000000'      ; enables peripheral interrupts
        movwf    INTCON           ;
        movlw    b'00000111'      ; turns off comparator
        movwf    CMCON            ;
        clrf     T1CON            ; turns off TMR1
        movlw    b'00001101'      ; turns on ADC, selects AN3,
        movwf    ADCON0           ;   relative to VDD, left-justified

        retfie                    ;

isr     movwf    W_temp           ; stores w. reg in temp register
        movfw    STATUS           ; stores STATUS in temporary
        movwf    STATUS_temp      ;   register

        bcf      STATUS, RP0      ; goes to Bank 0
        bcf      PIR1, 6          ; clears A/D interrupt flag

        bsf      STATUS, RP0      ; goes to Bank 1
        movlw    0x80             ; subtracts lower byte
        subwf    ADRESL, w        ;
        comf     STATUS, w        ; inverts carry flag (bit 0 of STATUS)
        andlw    b'00000001'      ; masks all other bits
        bcf      STATUS, RP0      ; goes to Bank 0
        addlw    0x12             ; add this to the number we are
        subwf    ADRESH, w        ;   subtracting from the higher byte
        btfss    STATUS, C        ;
        goto     Cold             ; ADRESH:L < 0x1280, so "cold!"

        bsf      STATUS, RP0      ; goes to Bank 1
        movlw    0x80             ; subtracts lower byte
        subwf    ADRESL, w        ;
        comf     STATUS, w        ; inverts carry flag (bit 0 of STATUS)
        andlw    b'00000001'      ; masks all other bits
        bcf      STATUS, RP0      ; goes to Bank 0
        addlw    0x15             ; add this to the number we are
        subwf    ADRESH, w        ;   subtracting from the higher byte
        btfss    STATUS, C        ;
        goto     OK               ; ADRESH:L < 0x1580, so OK
        goto     Hot              ; ADRESH:L = 0x1580,
                                  ;   so Hot!

Cold    movlw    b'000001'        ; turns on 'cold' LED
        movwf    GPIO             ;
        goto     prereturn        ;
```

```
OK          movlw    b'000010'        ; turns on 'OK' LED
            movwf    GPIO             ;
            goto     prereturn        ;

Hot         movlw    b'000100'        ; turns on 'Hot' LED
            movwf    GPIO             ;
            goto     prereturn        ;

prereturn   movfw    STATUS_temp      ; restores STATUS register to
            movwf    STATUS           ;   original value
            swapf    W_temp, f        ; restores working register to
            swapf    W_temp, w        ;   original value
            retfie                    ; returns, enabling GIE
```

```
;============
; Program Start
```

```
Start       call     Init             ; sets everything up
Main        bsf      ADCON0, 1        ; start A/D conversion
            goto     Main             ;

            END
```

Program Q

```
;***************************************
; written by: John Morton              *
; date: 14/03/05                       *
; version: 1.0                         *
; file saved as: gardenlights.asm      *
; for PIC12F675                        *
; clock frequency: Int. 4 MHz          *
;***************************************
```

```
; Program Description: Intelligent garden lights controller.

            list     P=12F675
            include  "c:\pic\P12F675.inc"

            __config _INTRC_OSC_NOCLKOUT & _WDT_OFF &
                     _PWRTE_ON & _MCLRE_ON & _BODEN_ON
                     & _CP_OFF & _CPD_OFF
```

```
Midnight    equ      20
Threshold   equ      21
Mark125     equ      22
Post125     equ      23
```

```
Post75          equ     24
FiveMins        equ     25
W_temp          equ     26
STATUS_temp     equ     27

                #define  summer         GPIO, 1
```

;============
; Declarations:

```
                org     0               ; first instruction to be
                                        ;   executed
                goto    Start           ;

                org     4               ; interrupt service routine
                goto    isr             ;
```

;============
; Subroutines:

```
Init            bsf     STATUS, RP0     ; goes to Bank 1
                movlw   b'001111'       ; GP5: lights, GP4: day/night
                                        ;   LED
                movwf   TRISIO          ; GP3: button, GP2: summer
                                        ;   switch
                                        ; GP0: analogue input
                clrf    WPU             ; no weak pull-ups
                movlw   b'10000111'     ; pull-ups disabled, TMR0
                                        ;   prescaled
                movwf   OPTION_REG      ; by maximum amount (256)

                clrf    PIE1            ; turns off peripheral
                                        ;   interrupts
                clrf    IOC             ; turns off interrupt on
                                        ;   change int.
                clrf    VRCON           ; turns off comparator V. ref.
                movlw   b'00110001'     ; AN0 is only analogue input
                movwf   ANSEL           ;   and analogue clock = RC
                call    3FFh            ; calls calibration address
                movwf   OSCCAL          ; moves w. reg into OSCCAL

                bcf     STATUS, RP0     ; back to Bank 0
                movlw   b'00000111'     ; turns off comparator
                movwf   CMCON           ;
                clrf    T1CON           ; turns off TMR1
                movlw   b'00000001'     ; turns on ADC, input: AN0
                movwf   ADCON0          ;   left justified
```

```
                clrf    GPIO                ; lights off, 'day' LED on
                movlw   b'00010000'         ; enables INT interrupt only
                movwf   INTCON              ;

                retfie                      ; returns, enabling interrupts
```

;==

```
isr             movwf   W_temp              ; stores w. reg in temp register
                movfw   STATUS              ; stores STATUS in temporary
                movwf   STATUS_temp         ;   register

                bcf     INTCON, 1           ; resets interrupt flag
                movfw   GPIO                ;
                xorlw   b'100000'           ; toggles state of lights
                movwf   GPIO                ;

                movfw   STATUS_temp         ; restores STATUS register to
                movwf   STATUS              ;   original value
                swapf   W_temp, f           ; restores working register to
                swapf   W_temp, w           ;   original value

                retfie                      ;
```

;==

```
ADconv          bsf     ADCON0, 1           ; starts AD conversion
                btfsc   ADCON0, 1           ; has it finished?
                goto    ADconv+1            ; no
                return                      ;
```

;==

```
Delay5min       movfw   TMR0                ; resets timing registers
                addlw   d'125'              ;
                movwf   Mark125             ;
                movlw   d'125'              ; sets up timing registers
                movwf   Post125             ;
                movlw   d'75'               ;
                movwf   Post75              ;

TimeLoop        movfw   Mark125             ; creates a five minute delay
                subwf   TMR0, w             ;
                btfss   STATUS, Z           ;
                goto    TimeLoop            ;
                movlw   d'125'              ;
                addwf   Mark125, f          ;
                decfsz  Post125, f          ;
                goto    TimeLoop            ;
                movlw   d'125'              ;
```

```
            movwf     Post125         ;
            decfsz    Post75, f       ;
            goto      TimeLoop        ;

            return                    ; 5 minutes have passed
```

;===
; Program Start

```
Start       call      Init            ; initialisation routine

            bsf       STATUS, RP0     ; Bank 1
            btfsc     PCON, 1         ; Power-Up or MCLR reset?
            goto      SetThreshold    ; MCLR reset
            bsf       PCON, 1         ; Power-up; resets POR bit

            clrf      EEADR           ;
            bsf       EECON1, 0       ; reads EEPROM address 0
            movfw     EEDATA          ; moves read data into w. reg
            movwf     Midnight        ;
            incf      EEADR           ;
            bsf       EECON1, 0       ; reads EEPROM address 1
            movfw     EEDATA          ;
            bcf       STATUS, RP0     ; Bank 0
            movwf     Threshold       ;
            goto      Main            ;

SetThreshold
            bcf       STATUS, RP0     ; Bank 0
            call      ADconv          ; perform A/D conversion
            movfw     ADRESH          ; takes 8 most significant bits
            movwf     Threshold       ;

            bsf       STATUS, RP0     ; Bank 1
            movwf     EEDATA          ; stores Threshold in EEPROM
            movlw     1               ; selects EEPROM address 1
            movwf     EEADR           ;
            bsf       EECON1, 2       ; enables a write operation
            bcf       INTCON, 7       ; disables global interrupts

            movlw     0x55            ; now follows the 'safe
            movwf     EECON2          ;   combination'
            movlw     0xAA            ;
            movwf     EECON2          ;
            bsf       EECON1, 1       ; starts the write operation
EELoop      btfsc     EECON1, 1       ; has write operation finished?
            goto      EELoop          ; no, so keeps looping
```

```
        bcf     STATUS, RP0     ; Bank 0
        bsf     INTCON, 7       ; re-enables global interrupts

        clrf    Midnight        ; resets Midnight register
        goto    Dusk            ;
;=================================================
Main    call    ADconv          ; this is the standard loop
        movfw   Threshold       ; is it Dusk?
        subwf   ADRESH, w       ;
        btfsc   STATUS, C       ;
        goto    Main            ; no

;====
Dusk    clrf    FiveMins        ; resets timing register
        movlw   b'110000'       ; turns on garden lights and
        movwf   GPIO            ;   'night' LED

Night   call    Delay5min       ; inserts 5 minute delay
        incf    FiveMins        ; counts up no. of 5 minutes

        movlw   d'12'           ; has 1 hour passed?
        subwf   FiveMins, w     ;
        btfss   STATUS, C       ;
        goto    Night           ; no

        movfw   Midnight        ; is it past midnight?
        subwf   FiveMins, w     ;
        btfss   STATUS, C       ;
        goto    Night           ; no

LightsOff movlw b'010000'       ; turns off garden lights and
        movwf   GPIO            ;   keeps 'night' LED on
        call    ADconv          ; performs A/D conversion
        movfw   Threshold       ; is it Dawn?
        subwf   ADRESH, w       ;
        btfss   STATUS, C       ;
        goto    Night           ; no

;====
Dawn    bsf     day             ; turns on 'day' LED
                                ; determines new midnight
        bcf     STATUS, C       ;
        rrf     FiveMins, w     ; divides time by 2
        btfss   summer          ; are we in summer time?
        sublw   d'24'           ; yes, subtracts 2 hours
        movwf   Midnight        ;
```

```
                bsf      STATUS, RP0    ; Bank 1
                movwf    EEDATA         ; stores Midnight in EEDATA
                clrf     EEADR          ; selects EEPROM address 00
                bsf      EECON1, 2      ; enables a write operation
                bcf      INTCON, 7      ; disables global interrupts

                movlw    0x55           ; now follows the 'safe
                movwf    EECON2         ;   combination'
                movlw    0xAA           ;
                movwf    EECON2         ;
                bsf      EECON1, 1      ; starts the write operation
EELoop2         btfsc    EECON1, 1      ; has write operation finished?
                goto     EELoop2        ; no, so keeps looping
                bcf      STATUS, RP0    ; Bank 0
                bsf      INTCON, 7      ; re-enables global interrupts
                clrf     FiveMins       ;

DawnLoop        call     Delay5min      ; one hour delay before looping back
                incf     FiveMins       ;
                movlw    d'12'          ; has one hour passed?
                subwf    FiveMins, w    ;
                btfss    STATUS, C      ;
                goto     DawnLoop       ; no
                goto     Main           ;

                END
```

Appendix A
Specifications of some Flash PIC microcontrollers

Device	Pins	I/O	Program memory	RAM	EEPROM	ADC	Other features
PIC10F200	6/8*	4	256	16	No	No	Internal 4 MHz oscillator, weak pull-ups, wake-up on change, 2-level stack
PIC10F202	6	4	512	24	No	No	As PIC10F202
PIC10F206	6	4	512	24	No	No	As PIC10F202, with comparator
PIC10F222	6	4	512	24	No	Yes	As PIC10F202
PIC12F508	8	6	512	25	No	No	As PIC10F202
PIC12F509	8	6	1024	41	No	No	As PIC12F508
PIC12F510	8	6	1024	38	No	Yes	As PIC12F508
PIC16F54	18	12	512	25	No	No	2-level stack
PIC16F57	28	20	2048	72	No	No	As PIC16F54
PIC16F59	40	32	2024	134	No	No	As PIC16F54
PIC16F84A	18	13	1024	64	64 bytes	No	8-level stack, interrupts
PIC12F675	8	6	1024	64	128 bytes	Yes	As PIC12F508, 16-bit TMR1, Comparator, 8-level stack, Interrupts
PIC16F676	14	12	1024	64	128 bytes	Yes	As PIC12F675

(Appendix A contd.)

(*Appendix A contd.*)

Device	Pins	I/O	Program memory	RAM	EEPROM	ADC	Other features
PIC16F627	8	16	1024	224	128 bytes	No	TMR1 (16-bit), TMR2 (8-bit), Comparator, 8-level stack, Interrupts, Capture/ Compare/PWM

Common features: All the PIC microcontrollers listed here have an 8-bit TMR0, a WDT (Watchdog timer), a DRT (device reset timer), POR (power-on reset), a lower power sleep mode, and support ICSP (In-circuit serial programming).

* The P10F2xx series have 6 pins in the surface mount package, but 8 pins in the larger packages (the two extra pins are N/C).

Appendix B
Pin layouts of some Flash
PIC microcontrollers

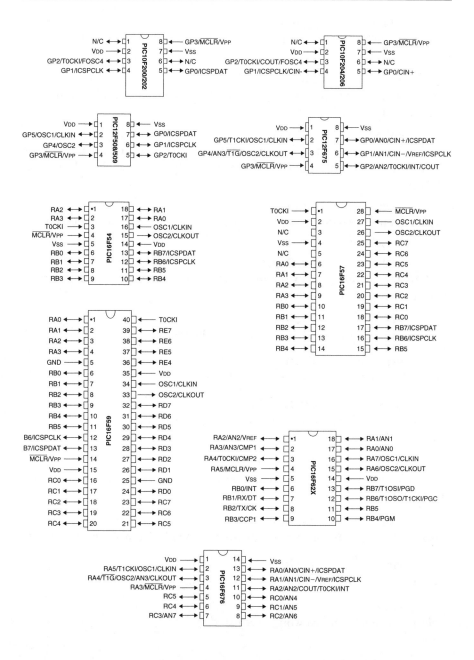

Appendix C
Instructions glossary

addlw   **number**
– (Not for PIC5x series) – **add**s a **number** with the number in the working register.

addwf   **FileReg, f**
– **add**s the number in the working register to the number in a file register and puts the result in the file register.

addwf   **FileReg, w**
– **add**s the number in the working register to the number in a file register and puts the result back into the working register, leaving the file register unchanged.

andlw   **number**
– **AND**s a **number** with the number in the working register, leaving the result in the working register.

andwf   **FileReg, f**
– **AND**s the number in the working register with the number in a file register and puts the result in the file register.

bcf   **FileReg, bit**
– **c**lears a **b**it in a file register (i.e. makes the bit 0).

bsf   **FileReg, bit**
– **s**ets a **b**it in a file register (i.e. makes the bit 1).

btfsc   **FileReg, bit**
– **t**ests a **b**it in a file register and **s**kips the next instruction if the result is **c**lear (i.e. if that bit is 0).

btfss   **FileReg, bit**
– **t**ests a **b**it in a file register and **s**kips the next instruction if the result is **s**et (i.e. if that bit is 1).

call   **AnySub**
– makes the chip **call** a subroutine, after which it will return to where it left off.

clrf   **FileReg**
– **clear**s (makes 0) the number in a file register.

clrw
– **clear**s the number in the working register.

clrwdt
– clears the number in the watchdog timer.

comf   **FileReg, f**
– complements (inverts, ones become zeroes, zeroes become ones) the number in a file register, leaving the result in the file register.

decf   **FileReg, f**
– decrements (subtracts one from) a file register and puts the result in the file register.

decfsz   **FileReg, f**
– decrements a file register and if the result is zero it skips the next instruction. The result is put in the file register.

goto   **Anywhere**
– makes the chip go to somewhere in the program which YOU have labelled 'Anywhere'.

incf   **FileReg, f**
– increments (adds one to) a file register and puts the result in the file register.

incfsz   **FileReg, f**
– increments a file register and if the result is zero it skips the next instruction. The result is put in the file register.

iorlw   **number**
– inclusive ORs a **number** with the number in the working register.

iorwf   **FileReg, f**
– inclusive ORs the number in the working register with the number in a file register and puts the result in the file register.

movfw   **FileReg**
or **movf**   **FileReg, w**
– moves (copies) the number in a file register in to the working register.

movlw   **number**
– moves (copies) a **number** into the working register.

movwf   **FileReg**
– moves (copies) the number in the working register into a file register.

nop
– this stands for: no operation, in other words – do nothing (it seems useless, but it's actually quite useful!).

option
– (Not to be used except in PIC5x series) – takes the number in the working register and moves it into the **option** register.

retfie
– (Not for PIC5x series) – returns from a subroutine and enables the Global Interrupt Enable bit.

retlw   number

– returns from a subroutine with a particular **number** (literal) in the working register.

return

– (Not for PIC5x series) – **return**s from a subroutine.

rlf   FileReg, f

– rotates the bits in a file register to the left, putting the result in the file register.

rrf   FileReg, f

– rotates the bits in a file register to the right, putting the result in the file register.

sleep

– sends the PIC to **sleep**, a lower power consumption mode.

sublw   number

– (Not for PIC5x series) – **sub**tracts the number in the working register from a **number**.

subwf   FileReg, f

– **sub**tracts the number in the working register from the number in a file register and puts the result in the file register.

swapf   FileReg, f

– **swap**s the two halves of the 8 bit binary number in a file register, leaving the result in the file register.

tris   PORTX

– (Not to be used except in PIC16C5x series) – uses the number in the working register to specify which bits of a port are inputs (correspond to a binary 1) and which are outputs (correspond to 0).

xorlw   number

– exclusive **OR**s a **number** with the number in the working register.

xorwf   FileReg, f

– exclusive **OR**s the number in the working register with the number in a file register and puts the result in the file register.

Appendix D
Number system conversion

	0	1	2	3	4	5	6	7	8	9	A	B	C	D	E	F
0	0	1	2	3	4	5	6	7	8	9	10	11	12	13	14	15
1	16	17	18	19	20	21	22	23	24	25	26	27	28	29	30	31
2	32	33	34	35	36	37	38	39	40	41	42	43	44	45	46	47
3	48	49	50	51	52	53	54	55	56	57	58	59	60	61	62	63
4	64	65	66	67	68	69	70	71	72	73	74	75	76	77	78	79
5	80	81	82	83	84	85	86	87	88	89	90	91	92	93	94	95
6	96	97	98	99	100	101	102	103	104	105	106	107	108	109	110	111
7	112	113	114	115	116	117	118	119	120	121	122	123	124	125	126	127
8	128	129	130	131	132	133	134	135	136	137	138	139	140	141	142	143
9	144	145	146	147	148	149	150	151	152	153	154	155	156	157	158	159
A	160	161	162	163	164	165	166	167	168	169	170	171	172	173	174	175
B	176	177	178	179	180	181	182	183	184	185	186	187	188	189	190	191
C	192	193	194	195	196	197	198	199	200	201	202	203	204	205	206	207
D	208	209	210	211	212	213	214	215	216	217	218	219	220	221	222	223
E	224	225	226	227	228	229	230	231	232	233	234	235	236	237	238	239
F	240	241	242	243	244	245	246	247	248	249	250	251	252	253	254	255

Appendix E
Bit assignments of various file registers

OPTION_REG

Bit no.	7	6	5	4	3	2	1	0
Bit name	...	...	T0CS	T0SE	PSA	PS2	PS1	PS0

Prescaler value...			TMR0	WDT
0	0	0	1:2	1:1
0	0	1	1:4	1:2
0	1	0	1:8	1:4
0	1	1	1:16	1:8
1	0	0	1:32	1:16
1	0	1	1:64	1:32
1	1	0	1:128	1:64
1	1	1	1:256	1:128

Prescaler Assignment
0: Prescaler assigned to TMR0
1: Prescaler assigned to WDT

TMR0 Source Edge Select
0: TMR0 counts on falling edge
1: TMR0 counts on rising edge

TMR0 Clock Source Select
0: TMR0 counts signal from oscillator
1: TMR0 counts signals on T0CKI pin

(PIC16F5X) **Unassigned**

(PIC12F5xx) **Weak Pull-ups Enable**
0: Enabled
1: Disabled

(PIC12F675 / PIC16F676 / F627) **Ext. Interrupt Edge Select**
0: Interrupt on falling edge of INT pin
1: Interrupt on rising edge of INT pin

(PIC16F5X) **Unassigned**

(PIC12F5xx) **Wake-up on Change Enable**
0: Enabled
1: Disabled

(PIC12F675 / PIC16F676 / PIC16F627) **Port B Pull-up Enable**
0: Weak pull-ups enabled on GPIO/Port B, if selected in WPU
1: Weak pull-ups disabled

STATUS

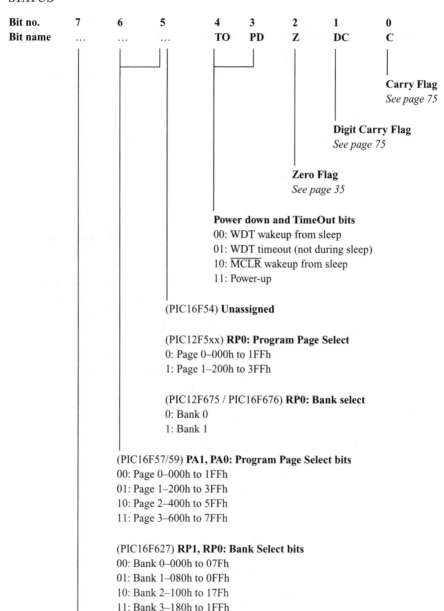

Bit no.	7	6	5	4	3	2	1	0
Bit name	...	...	...	TO	PD	Z	DC	C

Carry Flag
See page 75

Digit Carry Flag
See page 75

Zero Flag
See page 35

Power down and TimeOut bits
00: WDT wakeup from sleep
01: WDT timeout (not during sleep)
10: $\overline{\text{MCLR}}$ wakeup from sleep
11: Power-up

(PIC16F54) **Unassigned**

(PIC12F5xx) **RP0: Program Page Select**
0: Page 0–000h to 1FFh
1: Page 1–200h to 3FFh

(PIC12F675 / PIC16F676) **RP0: Bank select**
0: Bank 0
1: Bank 1

(PIC16F57/59) **PA1, PA0: Program Page Select bits**
00: Page 0–000h to 1FFh
01: Page 1–200h to 3FFh
10: Page 2–400h to 5FFh
11: Page 3–600h to 7FFh

(PIC16F627) **RP1, RP0: Bank Select bits**
00: Bank 0–000h to 07Fh
01: Bank 1–080h to 0FFh
10: Bank 2–100h to 17Fh
11: Bank 3–180h to 1FFh

(PIC16F627) **IRP: Indirect Addressing Bank Select**
0: Select Banks 0 and 1 for indirect addressing – 000h to 0FFh
1: Select Banks 2 and 3 for indirect addressing – 100h to 1FFh

INTCON

Bit no.	7	6	5	4	3	2	1	0
Bit name	GIE	PEIE	T0IE	INTE	GPIE	T0IF	INTF	GPIF

Port Change flag
0: It hasn't
[**Note:** Must be cleared by you]
1: GPIO/Port B change int. occurred

External INT flag
0: It hasn't
[**Note:** Must be cleared by you]
1: An interrupt has occurred on the INT pin

TMR0 Overflow Interrupt flag
0: TMR0 has not overflowed
[**Note:** Must be cleared by you]
1: TMR0 has overflowed

Port Change Interrupt Enable
0: Disables GPIO/Port B change interrupt
1: Enables it

External INT Interrupt Enable
0: Disables the INT pin interrupt
1: Enables it

TMR0 Overflow Interrupt Enable
0: Disables TMR0 overflow interrupt
1: Enables it

Peripheral Interrupt Enable
0: Disables any enabled 'peripheral interrupts'
1: Enables all 'peripheral interrupts'

Global Interrupt Enable
0: Disables ALL interrupts
1: Enables any enabled interrupts

PIE1/PIR1

The bit assignments in the peripheral interrupt registers PIE1 and PIR1 are identical. In PIE1 they refer to interrupt enable bits, and in PIR1 they refer to interrupt flags. The interrupt flags must be cleared by you in the interrupt service routine.

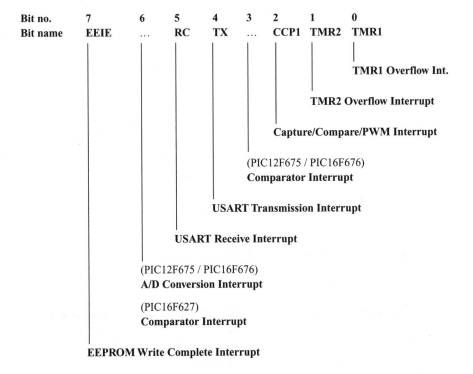

Bit no.	7	6	5	4	3	2	1	0
Bit name	EEIE	...	RC	TX	...	CCP1	TMR2	TMR1

TMR1 Overflow Int.

TMR2 Overflow Interrupt

Capture/Compare/PWM Interrupt

(PIC12F675 / PIC16F676)
Comparator Interrupt

USART Transmission Interrupt

USART Receive Interrupt

(PIC12F675 / PIC16F676)
A/D Conversion Interrupt

(PIC16F627)
Comparator Interrupt

EEPROM Write Complete Interrupt

PCON

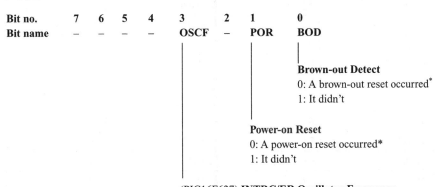

Bit no.	7	6	5	4	3	2	1	0
Bit name	–	–	–	–	OSCF	–	POR	BOD

Brown-out Detect
0: A brown-out reset occurred*
1: It didn't

Power-on Reset
0: A power-on reset occurred*
1: It didn't

(PIC16F627) **INTRC/ER Oscillator Frequency**
0: 37 kHz typical
1: 4 MHz typical

*Both these bits must be set in software, when cleared by the relevant reset.

EECON1

Bit no.	7, 6, 5, 4	3	2	1	0
Bit name	*unused*	WRERR	WREN	WR	RD

Read Control Bit
1: Starts an EEPROM read (gets cleared when read finishes)

Write Control Bit
1: Starts an EEPROM write operation (gets cleared when write finishes)

EEPROM Write Enable Bit
0: Forbids writing to the EEPROM
1: Permits writing to the EEPROM

EEPROM Write Error Flag
0: The write operation completed without error
1: An EEPROM write has prematurely terminated

VRCON

Bit 7: Comparator Voltage Reference Enable bit
0: Voltage reference module off (consuming no current)
1: Voltage reference module on

VRCON, 5 = 1 (low range)		VRCON, 5 = 0 (high range)	
VRCON 3:0	VRef (V_{DD} = 5V)	VRCON 3:0	VRef (V_{DD} = 5V)
0000	0.00	0000	1.25
0001	0.21	0001	1.41
0010	0.42	0010	1.56
0011	0.63	0011	1.72
0100	0.83	0100	1.88
0101	1.04	0101	2.03
0110	1.25	0110	2.19
0111	1.46	0111	2.34
1000	1.67	1000	2.50
1001	1.88	1001	2.66
1010	2.08	1010	2.81
1011	2.29	1011	2.97
1100	2.50	1100	3.13
1101	2.71	1101	3.28
1110	2.92	1110	3.44
1111	3.13	1111	3.59

CMCON

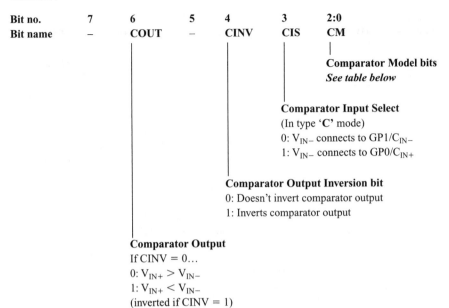

CMCON 2:0		V_{IN+}	V_{IN-}	V_{OUT}
000		GP0/C_{IN+}	GP1/C_{IN-}	*Disabled: CMCON, 6 = 0*
001	A	GP0/C_{IN+}	GP1/C_{IN-}	GP2/C_{OUT} *and* CMCON, 6
010		GP0/C_{IN+}	GP1/C_{IN-}	CMCON, 6
011	B	Internal ref.	GP1/C_{IN-}	GP2/C_{OUT} *and* CMCON, 6
100		Internal ref.	GP1/C_{IN-}	CMCON, 6
101	C	Internal ref.	GP0 *or* GP1	GP2/C_{OUT} *and* CMCON, 6
110		Internal ref.	GP0 *or* GP1	CMCON, 6
111		*Comparator off and consumes no current (CMCON, 6 = 0)*		

ADCON0

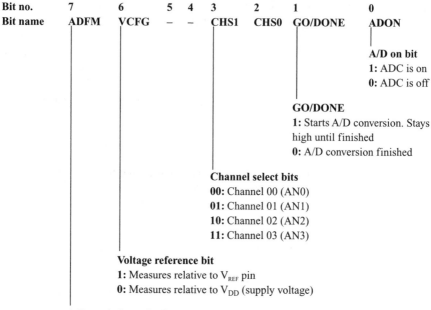

Bit no.	7	6	5	4	3	2	1	0
Bit name	ADFM	VCFG	–	–	CHS1	CHS0	GO/DONE	ADON

A/D on bit
1: ADC is on
0: ADC is off

GO/DONE
1: Starts A/D conversion. Stays high until finished
0: A/D conversion finished

Channel select bits
00: Channel 00 (AN0)
01: Channel 01 (AN1)
10: Channel 02 (AN2)
11: Channel 03 (AN3)

Voltage reference bit
1: Measures relative to V_{REF} pin
0: Measures relative to V_{DD} (supply voltage)

A/D result formed select
1: Right justified – result stored in ADRESL and ADRESH (bits 0:2)
0: Left justified – result stored in ADRESL (bits 6:7) and ADRESH

ANSEL

ANSEL bits 6:4	A/D conversion clock	Device frequency			
		1.25 MHz	**2.46 MHz**	**4 MHz**	**20 MHz**
000	$F_{OSC}/2$	1.6 μs	800 ns	500 ns	100 ns
001	$F_{OSC}/8$	6.4 μs	3.2 μs	2 μs	400 ns
010	$F_{OSC}/32$	25.6 μs	12.8 μs	8 μs	1.6 μs
011	F_{RC}: Internal oscillator	~4 μs	~4 μs	~4 μs	~4 μs
100	$F_{OSC}/4$	3.2 μs	1.6 μs	1 μs	200 ns
101	$F_{OSC}/16$	12.8 μs	6.4 μs	4 μs	800 ns
110	$F_{OSC}/64$	51.2 μs	25.6 μs	16 μs	3.2 μs
111	F_{RC}: Internal oscillator	~4 μs	~4 μs	~4 μs	~4 μs

ANSEL Bits 3:0 correspond to the four A/D input channels AN3:0.
0: Makes the pin a digital I/O pin
1: Makes the pin an analogue input, disabling weak pull-ups, interrupt-on-change, etc.

Appendix F
If all else fails, read this

You should find that there are certain mistakes which you make time and time again (I do!). I've listed the popular ones here:

1. Look for: **subwf**   **FileReg**   … are you sure you don't mean …

               subwf   **FileReg, w**

2. Have you remembered the correct addresses for general purpose file registers for your particular PIC model? (e.g. on the PIC12F675 they don't start until address **20h**).
3. *You are using a PIC microcontroller which has weak pull-ups* – have you remembered to set up bit 7 of the OPTION register correctly?
4. Are your subroutines in the correct page or half of page?
5. Are you adding something to the program counter in the wrong place on a page or on the wrong page?
6. Are you remembering to reset a file register you are using to keep track of how many times something has happened (e.g. a postscaler)?
7. *You think you are doing something to a file register but it isn't happening …* are you in the correct bank?
8. *If you are having a total nightmare and NOTHING is working …* have you specified the correct PIC microcontroller at the top of the program?
9. Have you set the configuration bits correctly when programming/simulating?

Appendix G
Contacts and further reading

John Morton:	help@to-pic.com
	www.to-pic.com
Microchip:	www.microchip.com
	Microchip Technology Inc.
	2355 W. Chandler Blvd.
	Chandler, AZ 85224-6199
	USA
Microchip UK Sales:	Phone: +44-118-921-5869
	Fax: +44-118-921-5820
PIC Press:	www.to-pic.com
Third-party products:	Olimex — www.olimex.com
	Spark Fun Electronics — www.sparkfun.com
	Taylec Ltd. — www.taylec.co.uk

Books and magazines

Electronic Systems by M. W. Brimicombe. Nelson (1985).
(A great text for general electronics)

Everyday and Practical Electronics
(A monthly magazine which normally has a PIC project or two)

PIC-robotics by John Iovine. Tab Books (2004)
PIC Microcontroller Project Book by John Iovine. Tab Books (2004)
(Two popular books on PIC robotics and PIC Basic)

Appendix H
PICKit™ 1 & BFMP Info

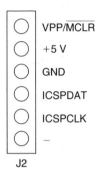

Figure H.1 *Pin assignment of the Baseline Flash Microcontroller Programmer.*

Programming a PIC microcontroller using the PICkit™ 1 Flash Starter Kit

To program a PIC microcontroller in MPLab, first load the .asm file and assemble it. Select the PICkit 1 programmer under Programmer → Select Programmer. To program the PIC microcontroller, go to Programmer → Program Device, or use the shortcut button. Set the configuration bits either using the **__config** command, or using the menu option in MPLab. Alternatively, generate a .hex file using your preferred method, and then use the custom PICkit 1 programming software (discussed in Chapter 2).

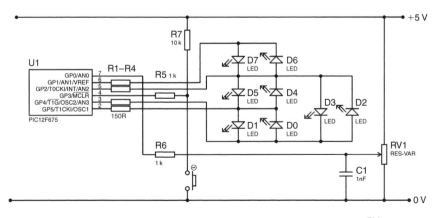

Figure H.2 *Components attached to the Evaluation Socket of the PICkit^{TM} 1 Flash Start Kit.*

Figure H.3 *Pin assignment of the jumper cable in the PICKit^{TM} 1 Flash Start Kit for the PIC12F675.*

Appendix I
Answers to the exercises

Chapter 1: Introduction

1.1 (a) Largest power of two less than $234 = 128 = 2^7$. Bit $7 = \mathbf{1}$

      This leaves $234 - 128 = 106$.   64 is less than 106 so bit $6 = \mathbf{1}$

      This leaves $106 - 64 = 42$.   32 is less than 42 so bit $5 = \mathbf{1}$

      This leaves $42 - 32 = 10$.   16 is greater than 10 so bit $4 = \mathbf{0}$

                                         8 is less than 10 so bit $3 = \mathbf{1}$

      This leaves $10 - 8 = 2$ 4 is greater than 2 so bit $2 = \mathbf{0}$

                                         2 equals 2 so bit $1 = \mathbf{1}$

      Nothing left so bit $0 = \mathbf{0}$

      The resulting binary number is **11101010**.

  (b) OR ...

      Divide 234 by two.   Leaves 117, remainder **0**

      Divide 117 by two.   Leaves 58, remainder **1**

      Divide 58 by two.   Leaves 29, remainder **0**

      Divide 29 by two.   Leaves 14, remainder **1**

      Divide 14 by two.   Leaves 7, remainder **0**

      Divide 7 by two.   Leaves 3, remainder **1**

      Divide 3 by two.   Leaves 1, remainder **1**

      Divide 1 by two.   Leaves 0, remainder **1**

      So **11101010** is the binary equivalent.

1.2 (a) Largest power of two less than $157 = 128 = 2^7$. Bit $7 = \mathbf{1}$

      This leaves $157 - 128 = 29$.   64 is greater than 29 so bit $6 = \mathbf{0}$

                                         32 is greater than 29 so bit $5 = \mathbf{0}$

                                         16 is less than 29 so bit $4 = \mathbf{1}$

      This leaves $29 - 16 = 13$.   8 is less than 13 so bit $3 = \mathbf{1}$

      This leaves $13 - 8 = 5$.   4 is less than 5 so bit $2 = \mathbf{1}$

      This leaves $5 - 4 = 1$.   2 is greater than 1 so bit $1 = \mathbf{0}$

                                         1 equals 1 so bit $0 = \mathbf{1}$

      The resulting binary number is **10011101**.

  (b) OR...

      Divide 157 by two.   Leaves 78, remainder **1**

      Divide 78 by two.   Leaves 39, remainder **0**

      Divide 39 by two.   Leaves 19, remainder **1**

      Divide 19 by two.   Leaves 9, remainder **1**

Divide 9 by two.   Leaves 4, remainder **1**
Divide 4 by two.   Leaves 2, remainder **0**
Divide 2 by two.   Leaves 1, remainder **0**
Divide 1 by two.   Leaves 0, remainder **1**

So **10011101** is the binary equivalent.

1.3   There are 14 16s in 234, leaving 234 − 224 = 10. So bit 1 = 14 = E, and bit 0 = 10 = A. The number is therefore **EA**.

1.4   There are 9 16s in 157, leaving 157 − 144 = 13. So bit 1 = 9, and bit 0 = 13 = D. The number is therefore **9D**.

1.5   1110 = 14 = E. 1010 = 10 = A. The number is therefore **EA**.

1.6   1. One push button requires **one** input.
       2. Four seven-segment displays require 4 + 7 = **11** outputs, creating a total of **12 I/O pins** which will just fit onto a **PIC54**.

1.7

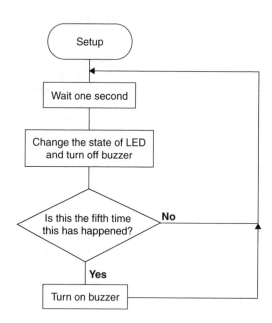

1.8   **b'0001'**   d'1'   1 h
       b'0010'   d'2'   2 h
       b'0100'   d'4'   4 h
       b'1000'   d'8'   8 h
       b'0001'   ...and so on.

1.9

```
Init    clrf     porta
        clrf     portb
        clrf     portc
        movlw    b'1001'
        tris     porta
        movlw    b'10000000'
        tris     portb
        movlw    b'00111110'
        tris     portc
        retlw    0
```

Chapter 2: Exploring the PIC16F5x series

2.1   Bits 6 and 7 are always 0.

The TMR0 is counting *externally*, so bit 5 is 1.

It's irrelevant whether the TMR0 is *rising* or *falling edge triggered* so bit 4 is 0 or 1 (let's say 0).

No prescaling for the TMR0 is required, so bit 3 is 1.

WDT is not be used, so WDT prescaling is irrelevant.

The number to be moved into the option register is **00101000**.

2.2   Bits 6 and 7 are always 0.

The TMR0 is counting *externally*, so bit 5 is 1.

It's irrelevant whether the TMR0 is *rising* or *falling edge triggered* so bit 4 is 0 or 1 (let's say 0).

Prescaling for the TMR0 is required, so bit 3 is 0.

$256 \times 4 = 1024$, so prescaling of 4 is required, so bit 2 is 0, bit 1 is 0, and bit 0 is 1.

The number to be moved into the option register is **00100001**.

2.3
```
        movlw    b'10101000'    ; moves the correct number into the
                                ;  working reg.
        xorwf    portb, f       ; toggles the correct bits in Port B
```

2.4   The **,f** or **,w** after the specified file register (e.g. **comf porta, f**) selects the destination of the instruction result. **,f** leaves the result in the file register and **,w** puts the result in the working register, leaving the file register unchanged.

2.5
```
        movlw    b'00010100'    ; motorists: green on, others off
        movwf    portb          ; pedestrians: red on, others off
```

2.6

ButtonLoop	btfss	**porta, 0**	; is the pedestrians' button
			; pressed?
	goto	**ButtonLoop**	; no, so loops back

2.7

	bsf	**portb, 1**	; turns motorists' amber light on
	bcf	**portb, 2**	; turns motorists' green light off
OR	movlw	**b'00010010'**	; motorists: amber on, others off
	movwf	**portb**	; pedestrians: red on, green off

2.8

TimeDelay	movwf	**PostX**	; sets up variable postscaler
	movlw	**d'240'**	; sets up fixed marker
	movwf	**Mark240**	;
TimeLoop	movfw	**Mark240**	; waits for TMR0 to count up
	subwf	**TMR0, w**	; 240 times
	btfss	**STATUS, Z**	;
	goto	**TimeLoop**	; hasn't, so keeps looping
	movlw	**d'240'**	; resets Mark240
	addwf	**Mark240, f**	;
	decfsz	**PostX, f**	; does this X times
	goto	**TimeLoop**	;
	retlw		; returns after required time

2.9

	movlw	**b'00100001'**	; motorists: red on, amber off
	movwf	**portb**	; pedestrians: green on, red off

2.10

	movlw	**d'80'**	; sends message of 8 seconds to sub
	call	**TimeDelay**	; creates delay of required time

2.11

	bsf	**portb, 1**	; turns on motorists' amber light
	bcf	**portb, 0**	; turns off motorists' red light
OR	movlw	**b'00100010'**	; motorists: red off, amber on
	movwf	**portb**	; pedestrians: green remains on

2.12

	movlw	**d'8'**	; sets up Counter8 with an initial
	movwf	**Counter8**	; value of 8
FlashLoop	movlw	**d'5'**	; sends message of 0.5 second
			; to sub
	call	**TimeDelay**	; creates delay of required time
	movlw	**b'00100010'**	; toggles the states of the lights
	xorwf	**portb, f**	;

```
            decfsz    Counter8, f   ; runs through this loop 8 times
            goto      FlashLoop     ;
```

2.13   dacgbfe

0	**11101110**
1	**00101000** or **00000110**
2	**11011010**
3	**11111000**
4	**00111100**
5	**11110100**
6	**11110110**
7	**01101000**
8	**11111110**
9	**11111100**
A	**01111110**
b	**10110110**
c	**10010010**
d	**10111010**
E	**11010110**
F	**01010110**

2.14

Clock cycle	*Instruction executed*	*PC*
1	0043	0044
2	–	0045
3	0045	0046
4	–	0048
5	0048	0049
6	0049	0050
7	–	0043
8	0043	0044 …

The cycle therefore repeats every 7 clock cycles.

2.15

```
Main   btfss    portb, 0    ; tests push button
       goto     Main        ; if not pressed, loops back
```

2.16

```
       incf     Counter, f  ;
```

2.17

```
       btfsc    Counter, 4  ; has Counter reached 16?
       clrf     Counter     ; if yes, resets Counter
```

2.18

```
       movfw    Counter     ; moves Counter into the
                            ;   working reg.
```

	call	_7SegDisp	; converts into 7-seg. code
	movwf	portb	; displays value
	goto	Main	; loops back to Main

2.19
_7SegDisp

	addwf	PCL	; skips a certain number of
			; instructions
	retlw	b'11111110'	; code for 0
	retlw	b'01100000'	; code for 1
	retlw	b'11011010'	; code for 2
	retlw	b'11110010'	; code for 3
	retlw	b'01100110'	; code for 4
	retlw	b'10110110'	; code for 5
	retlw	b'10111110	; code for 6
	retlw	b'11100000'	; code for 7
	retlw	b'11111110'	; code for 8
	retlw	b'11110110'	; code for 9
	retlw	b'11101110'	; code for A
	retlw	b'00111110'	; code for b
	retlw	b'10011100'	; code for C
	retlw	b'01111010'	; code for d
	retlw	b'10011110'	; code for E
	retlw	b'10001110'	; code for F

2.20
TestLoop	btfsc	portb, 0	; tests push button
	goto	TestLoop	; still pressed, so keeps looping
	goto	Main	; released, so returns

2.21
Delay	movlw	FFh	; adds 255 to TMR0, leaving result in
	addwf	TMR0, w	; the working register. Then moves
	movwf	Mark255	; the result into a marker register
TimeLoop	movfw	Mark255	; waits for the TMR0 to advance 255
	subwf	TMR0, w	;
	btfss	STATUS, Z	;
	goto	TimeLoop	; keeps looping
	retlw	0	; returns from sub after 0.07 second

2.22
Pressed

	call	Update	; updates timing and display
	btfss	bounce	; is button safe to test?
	goto	Pressed	;
	btfsc	portb, 0	; is button still pressed?

```
        goto    Pressed         ; yes, so loops
        call    Primebounce     ; activates de-bouncing routine
        goto    Released        ; loops back to 'Released' section
```

2.23   The number required would be **00000011**.

2.24
```
Display1    movfw   Seconds         ; takes the number out of
                                    ;   Seconds
            call    _7SegDisp       ; converts the number into 7-
                                    ;   seg. code
            movwf   portb           ; displays the value through
                                    ;   Port B

            movlw   b'0001'         ; turns on correct display
            movwf   porta           ;

            retlw   0               ; returns
Display10   movfw   TenSecond       ; takes the number out of
                                    ;   TenSecond
            call    _7SegDisp       ; converts the number into 7-
                                    ;   seg. code
            movwf   portb           ; displays the value through
                                    ;   Port B

            movlw   b'1000'         ; turns on correct display
            movwf   porta           ;

            retlw   0               ; returns
DisplayMin  movfw   Minutes         ; takes the number out of
                                    ;   Minutes
            call    _7SegDisp       ; converts the number into 7-
                                    ;   seg. code
            movwf   portb           ; displays the value through
                                    ;   Port B

            movlw   b'0100'         ; turns on correct display
            movwf   porta           ;
            retlw   0               ; returns
```

2.25
```
            movlw   d'10'           ; tests to see whether Seconds
            subwf   Seconds, w      ;   has reached 10 (i.e. whether
                                    ;   or not ten seconds have
            btfss   STATUS, Z       ;   passed)
            retlw   0               ; 10 seconds haven't passed, so
                                    ;   returns
```

```
             clrf      Seconds          ; 10 seconds have passed, so
             incf      TenSecond, f     ;   resets Seconds and
                                        ;   increments the number of
                                        ;   tens of seconds

             movlw     d'6'             ; tests to see whether
             subwf     TenSecond, w     ;   TenSecond has reached 6
                                        ;   (i.e. whether or not one
             btfss     STATUS, Z        ;   minute has passed)
             retlw     0                ; 1 minute hasn't passed, so
                                        ;   returns

             clrf      TenSecond        ; 1 minute has passed, so resets
             incf      Minutes, f       ;   TenSecond and increments
                                        ;   the number of minutes
```

2.26

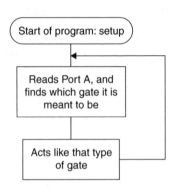

2.27   The resulting number would be **00010000**.

2.28
IORNOR

```
             movfw     porta            ; takes Input B
             iorwf     portb,w          ; IORs with Input A
             goto      Common           ; rest is as XOR/XNOR section
```

BUFFERNOT

```
             movfw     portb            ; takes Input A unchanged
             goto      Common           ; rest is as XOR/XNOR section
```

2.29
Main

```
             btfsc     porta, 1         ; tests setting switch
             goto      GreenLed         ; switch is high, so turn on
                                        ;   green LED
             bsf       portb, 1         ; switch is low, so turn on red
                                        ;   LED
```

2.30
GreenLed   bsf   portb, 0   ; turns on green LED

	goto	TenthSecond	; loops back to main body of
			; program

2.31

TenthSecond	movfw	TMR0	; is TMR0 at 0?
	btfss	STATUS, Z	;
	goto	Continue	;
	incf	TMR0	;
	decfsz	Post256	;
	goto	Continue	;
	clrf	portb	; it has, so turns off all LEDs
Continue	etc …		

2.32

	btfsc	porta, 1	; tests setting switch
	goto	TenthSecond	; disabled, so doesn't test
			; trigger input

2.33

	btfss	porta, 0	; tests to see whether motion
			; sensor has been set
	goto	TenthSecond	; not triggered, so loops back

2.34

	bsf	porta, 3	; turns on siren
EndLoop	clrwdt		; resets watchdog timer
	goto	EndLoop	; constantly loops

2.35

	bsf	STATUS, PA0	; selects Page 3
	bsf	STATUS, PA1	; selects Page 3
	goto	Earth	; now able to jump to Earth

2.36

Start	btfsc	STATUS, 4	; we need only test the
			; TimeOut bit
	call	PreInit	; set, so calls subroutine
	etc.		; clear, so skips subroutine

2.37

	bcf	FSR, 5	; selects GPFs 50-5F
	bsf	FSR, 6	; selects GPFs 50-5F
	movfw	Soldier	; copies number from Soldier
	bsf	FSR, 5	; selects GPFs 70-7F
	movwf	Spy	; copies number into Spy

Chapter 3: The PIC12F50x series

3.1

RandomScroll

	incf	Ran1, f	; quickly increments Ran1
			; & 2

```
            movlw    d'6'           ; has Ran1 reached 6?
            subwf    Ran1, w        ;
            btfss    STATUS, Z      ;
            retlw    0              ; no, so returns
            clrf     Ran1           ; yes, so resets Ran1
            incf     Ran2, f        ;
            movlw    d'6'           ; has Ran2 reached 6?
            subwf    Ran2, w        ;
            btfss    STATUS, Z      ;
            retlw    0              ; no, so returns
            clrf     Ran2           ; yes, so resets Ran2
            retlw    0              ;
```

3.2
RandomGen

```
            movlw    d'63'          ; new Random =
            addwf    Random, w      ;   63 + old Random x 3
            addwf    Random, w      ;
            addwf    Random, f      ;
            retlw    0              ;
```

3.3
Slowdown

```
            incf     PostVal, f     ; increments PostVal until
            btfsc    PostVal, 5     ;   it reaches 32, upon which
            clrf     PostVal        ;   it is reset to 0
            retlw    0              ;
```

3.4
Display

```
Display     btfss    TMR0, 4        ; uses bit 4 of TMR0 to choose
            goto     Die2           ;   which die
            movfw    Die1num        ; gets number to display
            call     Code1          ; converts to code
            movwf    GPIO           ; outputs
            retlw    0              ;

Die2        movfw    Die2num        ; gets number to display
            call     Code2          ; converts to code
            movwf    GPIO           ; outputs
            retlw    0              ;
```

; pin arrangement is: CTLR, A, -, B, C, D for GPIO 5:0

```
Code1       addwf    PCL, f         ;
            retlw    b'100000'      ; all off
            retlw    b'110000'      ; 1
            retlw    b'100100'      ; 2
            retlw    b'110100'      ; 3
            retlw    b'100110'      ; 4
            retlw    b'110110'      ; 5
```

```
          retlw    b'100111'        ; 6
          retlw    b'110111'        ; all on
Code2     addwf    PCL, f           ;
          retlw    b'010111'        ; all off
          retlw    b'000111'        ; 1
          retlw    b'010011'        ; 2
          retlw    b'000011'        ; 3
          retlw    b'010001'        ; 4
          retlw    b'000001'        ; 5
          retlw    b'010000'        ; 6
          retlw    b'000000'        ; all on
```

Chapter 4: Intermediate operations using the PIC12F675

```
4.1       bsf      STATUS, RP0      ; goes to Bank 1
          bsf      OPTION_REG, 6    ; selects rising edge INT trigger
          bcf      STATUS, RP0      ; back to Bank 0
          movlw    b'10010000'      ; enables INT and global interrupts
          movwf    INTCON           ;
          sleep                     ; goes to sleep
          nop                       ; this line is executed upon wake-up,
                                    ;   but it does nothing

4.2
Init      bsf      STATUS, RP0      ; goes to Bank 1
          call     3FFh             ; calls calibration address
          movwf    OSCCAL           ; moves w. reg into OSCCAL

          movlw    b'011110'        ; GP5: Buzzer, GP3: Reset button
          movwf    TRISIO           ; GP1,2,4: LEDs/Buttons (inputs to
                                    ;   start with), GP0: LED enable
          movlw    b'010110'        ; GP1,2,4 have weak pull-ups
          movwf    WPU              ;   enabled

          movlw    b'00000111'      ; pull-ups enabled, TMR0 prescaled
          movwf    OPTION_REG       ;   by maximum amount (256)
          clrf     PIE1             ; turns off peripheral interrupts
          movlw    b'010110'        ; enables GPIO change interrupt on
          movwf    IOC              ;   GP1, GP2 and GP4 only
          clrf     VRCON            ; turns off comparator V. ref.
          clrf     ANSEL            ; makes GP0:3 digital I/O pins

          bcf      STATUS, RP0      ; back to Bank 0
          clrf     GPIO             ; resets input/output port
          movlw    b'00001000'      ; enables GPIO change interrupt
                                    ;   only
          movwf    INTCON           ;
          movlw    b'00000111'      ; turns off comparator
```

	movwf	CMCON	;
	clrf	T1CON	; turns off TMR1
	clrf	ADCON0	; turns off A to D converter
	retfie		; returns, enabling interrupts
4.3	btfss	INTCON, 0	; checks GPIO interrupt flag
	goto	Timer	; TMR0 interrupt occurred…
			; GPIO interrupt occurred…
4.4	bsf	STATUS, RP0	; moves to Bank 1
	movlw	b'001000'	; makes GP1,2,4 outputs
	movwf	TRISIO	;
	bcf	STATUS, RP0	; moves to Bank 0
	movfw	temp	; moves temp back into GPIO,
	addlw	b'100001'	; sets GP5 and GP0 (turns on
	movwf	GPIO	; buzzer and enables LEDs)
4.5	movlw	b'00100000'	; enables TMR0 interrupt, disables
	movwf	INTCON	; the GPIO change interrupt
	retfie		; returns, enabling GIE

4.6

Timer	bcf	INTCON, 2	; resets TMR0 interrupt flag
	decfsz	Post16, f	; is this the 16th TMR0 interrupt
	retfie		;
	bcf	GPIO, 5	; turn off buzzer
	clrf	INTCON	; turns off all interrupts
	sleep		; goes into low power mode
4.7	bsf	STATUS, RP0	; goes to Bank 1
	movlw	08h	; moves the address to be read (08h)
	movwf	EEADR	; into EEADR
	bsf	EECON1, 0	; reads EEPROM
	movlw	d'5'	; adds 5 to the value which
	addwf	EEDATA, f	; was read
	incf	EEADR	; address to be written to is 09h
	bsf	EECON1, 2	; enables a write operation
	bcf	INTCON, 7	; disables global interrupts
	movlw	55h	; now follows the 'safe combination'
	movwf	EECON2	;
	movlw	AAh	;
	movwf	EECON2	;
	bsf	EECON1, 1	; starts the write operation

EELoop			
	btfsc	EECON1, 1	; has write operation finished?
	goto	EELoop	; no, still high, so keeps looping

4.8   **INTCON:**   **b'00010000'**
(only the INT interrupt is enabled – don't enable global yet)

TRISIO: **b'00011110'**
(GP0 is the only output)

WPU: **b'00000000'**
(not used)

OPTION_REG: **b'11000111'**
(for counting minutes, prescale TMR0 by maximum)

4.9

Main			
	bsf	STATUS, RP0	; selects Bank 1
	clrf	EEADR	; selects EEPROM address 00h
	bsf	EECON1, 0	; initiates an EEPROM read
	movfw	EEDATA	; reads EEDATA
	bcf	STATUS, RP0	; selects Bank 0
	btfss	STATUS, Z	; is it 0?
	goto	Active	; no, so goes to Active
	bcf	GPIO, 0	; turns off GP0
	sleep		; goes to sleep
	nop		;
	goto	Main	; loops back to Main

4.10

Active			
	bsf	GPIO, 0	; card has minutes
	btfss	GPIO, 1	; is a call in progress?
	goto	Active+1	; no, so keeps waiting
	movfw	Mark125	; has one minute passed?
	subwf	TMR0, w	;
	btfss	STATUS, Z	;
	goto	Active+1	; no, so keeps looping
	movlw	d'125'	;
	addwf	Mark125	;
	decfsz	Post125	;
	goto	Active+1	; no, so keeps looping
	movlw	d'125'	;
	movwf	Post125	;
	decfsz	Post15	;
	goto	Active	; no, so keeps looping

4.11	movlw	d'15'	; resets final postscaler
	movwf	Post15	;
	bsf	STATUS, RP0	; goes to Bank 1
	clrf	EEADR	; selects EEPROM address 00h
	bsf	EECON1, 0	; reads EEPROM address 00h

```
              decf       EEDATA              ; subtracts 1 minute from card

              bsf        EECON1, 2           ; enables a write operation
              bcf        INTCON, 7           ; disables global interrupts

              movlw      55h                 ; now follows the 'safe
              movwf      EECON2              ;   combination'
              movlw      AAh                 ;
              movwf      EECON2              ;
              bsf        EECON1, 1           ; starts the write operation
              bsf        INTCON, 7           ; enables global interrupts

EELoop        btfsc      EECON1, 1           ; has write operation finished?
              goto       EELoop              ; no, still high, so keeps looping
              bcf        STATUS, RP0         ; back to Bank 0
              goto       Main                ; loops back to start


4.12          bcf        INTCON, 1           ; resets INT interrupt flag
              bcf        STATUS, RP0         ; makes sure we're in Bank 0
              movfw      GPIO                ; reads value of GPIO
              movwf      temp                ;
              rrf        temp, f             ; rotates right three times...
              rrf        temp, f             ;
              rrf        temp, w             ; ...leaving result in w. reg
              andlw      b'000111'           ; masks bits 3–5


4.13
CardValue     addwf      PCL, f              ; returns with new number of
              retlw      d'2'                ;   minutes for the card
              retlw      d'5'                ;
              retlw      d'10'               ;
              retlw      d'20'               ;
              retlw      d'40'               ;
              retlw      d'60'               ; one hour
              retlw      d'120'              ; two hours
              retlw      0                   ; (erases card)
```

4.14   **INTCON:**   b'01000000'
          (peripheral interrupts enabled – don't enable global yet)

          PIE1:   b'01000000'
          (enables A/D conversion interrupt)

          TRISIO: b'010000'
          (GP0:2 are LEDs, GP4 is an analogue input, GP3 and 5
          are unused)

          WPU: 0   (weak pull-ups are off)

OPTION_REG: b'10000000'
(Weak pull-ups disabled. No timing functions are used)

ADCON0:   b'00001101'
(Turns on ADC. Selects channel AN3, relative to V_{DD}. Left-justified answer)

ANSEL: b'00011000'
(A/D clock: Fosc/8 = 2 μs; AN3 (GP4) is analogue input, others are digital)

4.15   **bcf   STATUS, RP0   ; selects Bank 0**
       bcf   PIR1, 6   ; clears A/D interrupt flag

4.16   42°C = 0.42 V
       0.42 V / 5 V = 0.084
       0.084 × 1024 = 86

       d'86' = b'00000010101 1000000' = 0x1580

This translates to **0x15** in ADRESH and **0x80** in ADRESL

4.17   **bsf   STATUS, RP0   ; goes to Bank 1**
       movlw   0x80   ; subtracts lower byte
       subwf   ADRESL, w   ;
       comf   STATUS, w   ; inverts carry flag (bit
                                          0 of STATUS)
       andlw   b'00000001'   ; masks all other bits
       bcf   STATUS, RP0   ; (goes to Bank 0)
       addlw   0x15   ; add this to the number we are
       subwf   ADRESH, w   ;   subtracting from the higher
                                          byte
       btfss   STATUS, C   ;
       goto   OK   ; ADRESH:L < 0x1580,
                                          ;   →OK
       goto   Hot   ; ADRESH:L ≥ 0x1580,
                                          ;   →Hot!

4.18
Cold movlw   b'000001'   ; turns on 'cold' LED
       movwf   GPIO   ;
       goto   prereturn   ;

OK   movlw   b'000010'   ; turn on 'OK' LED
       movwf   GPIO   ;
       goto   prereturn   ;

Hot   movlw   b'000100'   ; turn on 'Hot' LED
       movwf   GPIO   ;
       goto   prereturn   ;

Appendix J
Some BASIC commands
in assembly

Some users will be familiar with BASIC programming, indeed there are several PIC microcontroller development kits in which programs are written in BASIC, and then converted into assembly language. This conversion process can be very inefficient, and so naturally I would recommend writing PIC programs directly in assembly. To assist those with a background in BASIC programming, I have provided a table showing how to write some BASIC operations in assembly.

BASIC	Assembly language		
GOTO MAIN		goto	MAIN
GOSUB INIT		call	Init
RETURN		retlw	0
LET X = 9		movlw	d'9'
		movwf	X
LET X = X + 1		incf	X, f
LET X = X + 10		movlw	d'10
		addwf	X, f
LET X = Y		movfw	Y
		movwf	X
IF X = 10 THEN		movlw	d'10'
GOTO ARM		subwf	X, w
ELSE		btfsc	STATUS, Z
GOTO DISARM		goto	ARM
END IF		goto	DISARM
FOR X = 1 TO 30		movlw	d'1'
. . . ' add stuff here		movwf	X
NEXT X	Loop	. . .	; add stuff here
		incf	X, f
		movlw	d'31'
		subwf	X, w
		btfss	STATUS, Z
		goto	Loop
DO	Loop	. . .	; add stuff here
. . . ' add stuff here		movlw	d'10'
WHILE (X < 10)		subwf	X, w
		btfss	STATUS, C
		goto	Loop

Index